STRUCTURE AND BONDING

Volume 41

With 57 Figures and 3 Tables

Springer-Verlag
Berlin Heidelberg New York 1980

Editor of this volume:

Professor Dr. Peter Hemmerich
Universität Konstanz
Fachbereich Biologie
Postfach 333
D-7750 Konstanz

ISBN 3-540-09958-1 Springer-Verlag Berlin Heidelberg New York
ISBN 0-387-09958-1 Springer-Verlag New York Heidelberg Berlin

Library of Congress Catalog Card Number 67-11280

Typesetting: R. & J. Blank, München.
Printing and bookbinding: Brühlsche Universitätsdruckerei, Gießen
2152/3140-543210

SPRINGER-VERLAG

D-6900 Heidelberg 1
P. O. Box 105280
Telephone (06221) 487·1
Telex 04-61723

D-1000 Berlin 33
Heidelberger Platz 3
Telephone (030) 822001
Telex 01-83319

SPRINGER-VERLAG
NEW YORK INC.

175, Fifth Avenue
New York, N.Y. 10010
Telephone (212) 477-8200

Foreword

"Molecular Sensory Physiology" is the youngest among the biological subdisciplines, which are devoted to the understanding of biological activities in terms of chemical structure and bonding. In contrast to metabolic functions, which are concerned with a turnover or transfer of matter, and be it matter as small in size and weight as lone electrons, the new discipline is characterized by the transfer of "signals", the question being: What are the signals made of and how are they transmitted? The present volume is concerned with light signals and their transformation. The first step is the localization and identification of the light acceptor. The difficulties connected with this aim are mostly pertinent to the fact that signal transfer activity can only be followed in the whole organism, and only in rare cases on a subcellular level. Furthermore, the acceptor molecules (chromophores) are frequently not unique or specific in their signal transfer activity, but catalyze known metabolic functions in other places (environments). Thus, we must distinguish the possibilities of finding known holoproteins with hitherto unassigned functions or, on the other hand, known cofactors exhibiting new functions when bound by a new protein. Third, we may have in the new function a slightly modified cofactor. In any case, since we cannot follow the signal transfer activity down to a cell-free system, we must reverse our way of thinking and feed thc whole organism with a modified acceptor in order to trace down the signal transformation into "response" and, subsequently, explain the response in terms of chemical structure. In other words: The routine of "natural product chemistry" and its refined methods of structural microanalysis might not apply. Instead, new ways of "bioorganic" imagination applied to suitable mutants of active organisms might lead to success. The difficulties in betraying whole cells with a modified acceptor are enormous, but can be overcome by an intense cooperation between biologists and organic chemists. This cooperation is hampered by "linguistic" problems, which this volume might help to overcome.

P. Hemmerich

Contents

Physiological Bluelight Reception

Werner Schmidt*

Universität Konstanz, Fachbereich Biologie, Postfach 7733, D-7750 Konstanz, Federal Republic of Germany

Table of Contents

* This work was supported by the Deutsche Forschungsgemeinschaft (Sonderforschungsbereich 138, Projekt A1, "Biologische Grenzflächen und Spezifität").

W. Schmidt

This article surveys exemplary the great diversity of physiological bluelight effects including directional, metabolic and morphological responses. Emphasis is laid on the elucidation of the chemical nature, localization and orientation of the bluelight photoreceptor and the possible molecular mechanisms involved in the bluelight reception. Spectroscopical data, inhibitor- and photophysiological experiments are described, favoring a flavin rather than a carotenoid photoreceptor. Latest experiments regarding the usefullness of the cyt b -photoreduction as possible bluelight receptor assay are discussed, including the association of this very cyt b with the plasmamembrane. Finally, recent experiments on artificial membrane/flavin systems are taken into consideration, which offer the possibility to study the translation of light into a biochemical equivalent termed *sensory transduction* which finally leads to the physiological response.

1 Introduction

Life is indispensably dependent on visible light. Therefore, virtually all living systems have developed light-receiving apparatus and mechanisms, most probably derived from molecules and structures already otherwise employed. These photoreceptor systems, whether primitive or highly sophisticated, utilize specific pigments that transduce light, which may be used either as a source of energy or information, or both, into a biochemical equivalent via a relaxation process called *sensory transduction.*

Using the general development of molecular biology as a guide, we can assume that during phylogenetic development light reception and recognition has been optimized by a strong selectional pressure, modified somewhat by the type of niche occupied by the specific species. The energy of sunlight reaching the surface of the earth forms a broad maximum in the blue-green region of the spectrum[90]. Therefore, it is not surprising that – besides the commonly known predominant role of light in photosynthesis – the physiological effectiveness of bluelight is quite common throughout the living world. Countless examples of these phenomena indicate a variety of apparent advantages to the light-responsive organisms. For example, in potentially green organisms, the benefit involves ongoing or future photosynthesis. In certain molds, the spores are carried in balls at the end of the growing stalks called *sporangiophores*; these grow towards the light, and the widespread dispersal of spores is thereby enhanced. It seems reasonable to predict that regardless of the great variety of photobiological responses and peculiarities, all the phenomena can finally be reduced to a few basic principles, and that there are not much more physiological light receptors left to be discovered than those already known (carotenoids, flavins, rhodopsins, chlorophylls, phytochrome and a few others).

For bluelight reception, only flavins and carotenoids remain as serious contenders. However, as will be discussed in more detail below, more recent developments seem to favor the flavins, rather than the carotenoids, as photoreceptors and transducer components. Due to the extended field of physiological bluelight responses and recently developed analytical procedures (e.g., use of *photomutants*), it is appropriate to separately review bluelight research on *Phycomyces* (photodifferentiation and light growth response) and *Euglena* (phototaxis), see Russo/Galland and Doughty/Diehn (this volume).

As mentioned above, two of the central problems of biological bluelight reception are that of the photoreceptor pigment itself, and the subsequent sensory transduction mechanism. Obviously, the latter process is closely connected to the question of receptor molecule localization. As soon as this problem is elucidated for one specific organism and response, this can probably be extrapolated to other organisms and bluelight responses with only minor modifications. Therefore, it appears to be of secondary importance to know which of the many known bluelight responses are being investigated in which organism. However, it is not certain how far the concepts derived from those experimental restrictions, which are mostly quite artificial (such as use of etiolated plants, short exposure times, monochromatic light) can be extended to cover behavior under natural conditions.

No attempt is made to either provide an exhaustive coverage of literature in the bluelight receptor field, nor is it intended that specific organisms and responses be discussed in detail. For this purpose there are excellent reviews available covering the field of phototropism: DuBuy and Nuernbergk [52], Boysen-Jensen [19], Went and Thimann [189], van Overbeek [182], Schrank [158], Brauner [22], Went [190], Galston [64], Reinert [144], Thimann and Curry [177,178], Briggs [24,23], and Ball [4]. Photoresponses in fungi have been reviewed by Page [127] and Carlile [33]. An extensive, general review on *Phycomyces*, including sensory physiology with respect to bluelight, was written collectively by several authors [10]. During preparation of this manuscript, Presti and Delbrück published a general review article covering all important biological photoreceptors, briefly surveying physiological bluelight reception as well [139].

The present article is subdivided by means of phenomena: First, phototropism, which is the oldest topic in bluelight receptor research, is discussed in detail. Second, metabolic responses such as bluelight controlled carotenogenesis and oxygen exchange, and morphological responses, such as bluelight dependent rhythms and chloroplast movements, are reviewed. Effects of polarized light are taken into particular consideration, because they provide unique information on the localization and orientation of the bluelight receiving pigments not obtainable with other experimental techniques. Third, the current molecular approaches, such as bluelight induced photoreduction of a b-type cytochrome observed in several organisma, and its localization within the cell, are discussed. Finally, initial experiments on completely artificial flavin/membrane systems are presented, which might resemble the natural bluelight photoreceptor and its mechanism in sensory transduction.

For the purpose of comparison all spectra taken from the literature were digitized by means of a Hewlett-Packard Digitizer (model 9864 A) on-line with a small computer (HP, model 9820) and replotted on a common scale (HP, model 9862 A).

2 Phototropism

2.1 The Phenomena

The scientific research on this subject was founded in 1880 by the Darwins and published in their book *Power of Movements in Plants* [42]. Phototropic movements are restricted to the growing part of higher plants, such as young stems, leaves and flowers (Fig. 1). Surprisingly, in some cases even roots show directional responses [20]. Directional photoresponses are found in chloronemata of ferns [57], sporophores of mosses [171], and also in freely swimming organisms, including flagellata [80] and even bacteria [84]. However, most of the analytical work on this subject has been carried out on only a few species, such as grass coleoptiles, sporangiophores of fungi and on the green algae *Euglena*.

For directional responses, the photoreceptor molecules (or an array of those) are necessarily – at least for the duration of the photoreaction – located in a matrix with a fixed orientation to the bending organelle, or to the whole body of freely swimming organisms.

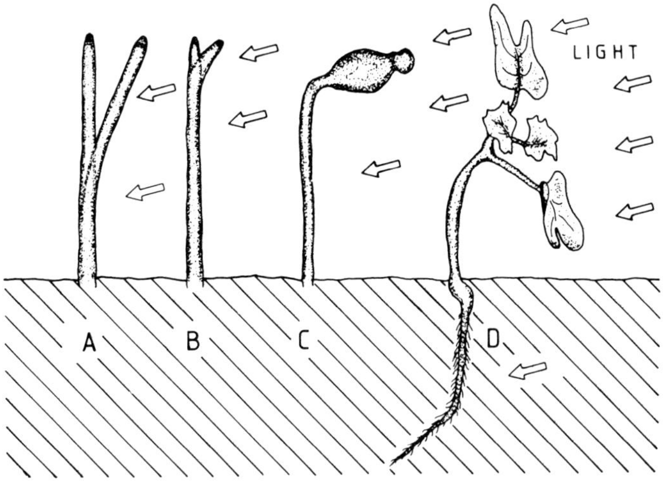

Fig. 1. Typical phototropic curvature as exhibited by (*A*) the maize coleoptile (second positive curbature), (*B*) the maize coleoptile (first positive curvature) and (*C*) the sporangiophore of the fungus *Philobolus*. (*D*): the hypocotyl of the mustard seedling bends towards the light (positive phototropism), the leaves orient their surface towards the light (diaphototropism), and the root bends away from the light (negative phototropism), according to Mohr[120)]. In all cases the response is only induced by bluelight

2.2 The Dosage Response Relationship

The most powerful tool in photobiology for elucidating the nature of a photoreceptor is to determine the action spectrum, which reflects – at least in its gross structure – the absorption spectrum of the photoreceptor pigment (for details see the article on action spectroscopy by Shropshire[164)]). In order to obtain a "true" action spectrum, dosage response curves are required. With respect to phototropic bending, the most accurate ones have been worked out by Zimmerman and Briggs[197)] for the avena coleoptile. These curves surprisingly exhibit complicated patterns. Up to a dosage of 10 pE/cm^2 of the stimulating bluelight bending follows the *Weber-Fechner* as well as the *Bunsen-Roscoe* law[1]: *First positive curvature*. Higher dosages result in smaller and even negative bending: *first negative curvature*. At still higher dosages a *second positive curvature* arises.

Redlight pretreatment decreases the phototropic sensitivity for the first positive curvature by a factor of 10. Con and Briggs[37)] demonstrated by action spectroscopy that this decrease is mediated by phytochrome. It appears that there is an interaction

1 The *Weber-Fechner law* is quite common in biology stating that physiological responses depend on the logarithm of the stimuli; the *Bunsen-Roscoe law* states that the action of photochemically active light depends only on the dosage d = I × t, regardless of how the single factors I (intensity) and t (time) are chosen.

of two different photoreceptor-response pathways. However, the decreased bluelight sensitivity of redlight pretreated seedlings may simply be due to phytochrome-induced production of other pigments, which, in turn, weaken the phototropically active bluelight. Absorption spectroscopy would elucidate this problem.

The complete dosage response data as obtained by Zimmerman and Briggs are replotted in Fig. 2. However, the abscissa is chosen to be a time scale rather than dosage, which demonstrates that for illumination stimuli lasting longer than about 15 min, it is only the time factor that counts, regardless of redlight pretreatment or light intensities employed. Such a situation is probably obtained under natural conditions in daylight.

Zimmerman and Briggs explain their dosage response curves on the basis of three independent pigment systems. However, for several reasons it appears more reasonable to ascribe their complicated patterns to different *secondary* rather than to distinct primary processes. First, the first and second positive curvatures show essentially the same action spectra (Fig. 3:4 and 5). Second, the *Bunsen-Roscoe* law holds only for the first 100 s of irradiation. After that time factors other than photochemical ones clearly govern phototropism. Third, the dosage response curves are *not* real kinetics, i.e. they do not represent continuous traces of bending in time, as the authors assume for their calculations. However, curvature was allowed to develop for 100 min in darkness, measured and plotted as a function of dosage.

For the analysis of the primary bluelight receptor action, it is necessary to uncouple the sensory transduction process from the subsequent phototropic reaction

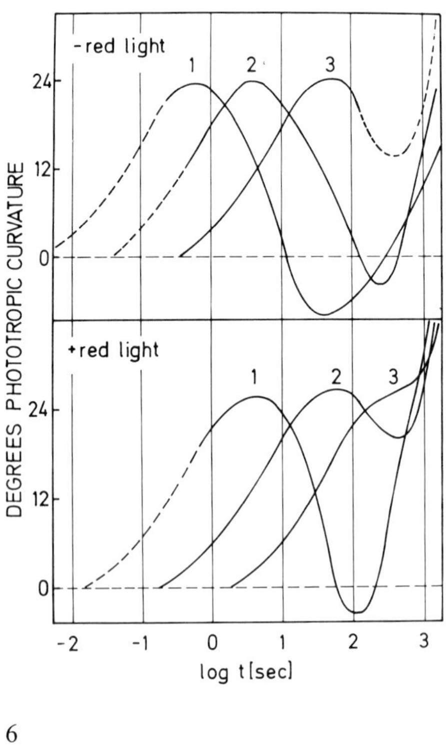

Fig. 2. Phototropic dosage response curves for oat coleoptiles at three intensities of blue light (440 nm): (*1*) $1.4 \cdot 10^{-11}$, (*2*) $1.4 \cdot 10^{-12}$ and (*3*) $1.4 \cdot 10^{-13}$ einsteins/-cm^2 s, with and without red light pretreatment. In contrast to the original figure[197] these curves are replotted on a common timescale (not dosage). After 100 s of bluelight irradiation all curves approach each other assymptotically, regardless of light intensities and light-pretreatment (see text)

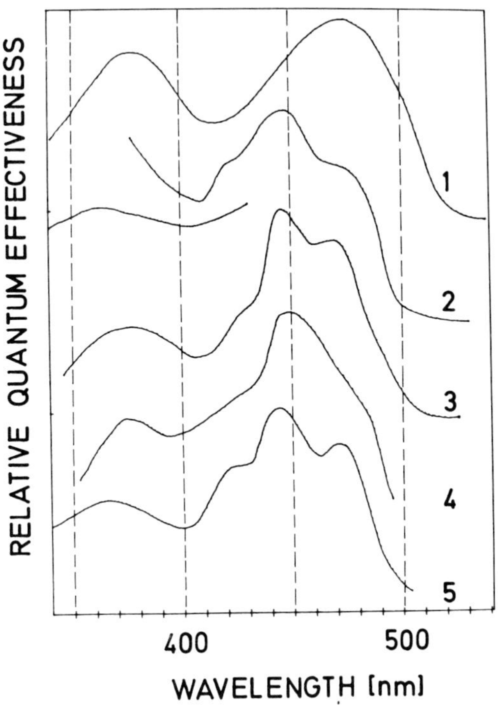

Fig. 3. Comparison of different bluelight action spectra for directional responses: (*1*) positive phototopotaxis of *Euglena*[48)], (*2*) Phototropism of *Philobolus*[127)], (*3*) phototropism of *Phycomyces*[40)], (*4*) second positive curvature of the avena coleoptile[58)], (*5*) first positive curvature of the avena coleotpile[39)]

chain. This was accomplished for the first time by v. Guttenberg[75)]. He irradiated oat coleoptiles at 5 °C for two hours from one side and kept the plants at the same temperature for another 10 h in darkness, without observing any bending. However, when transferred after the incubation time of 10 h to a warmer temperature, bending revived.

2.3 The Spectral Characteristics

2.3.1 Action Spectra

The shape of the different bluelight action spectra is by far not as uniform among the different organisms and responses as usually implied. Indeed, there is great diversity, especially with respect to the morphogenetic data (cf., Figs. 3, 16, 14, 23). The two individual action spectra on the phototropism of the avena coleoptile by Curry (Fig. 3 : 5,[39)]) and of the light-growth-response of *Phycomyces* by Curry and Gruen (Fig. 3 : 3,[40)]), therefore, appear to the present author to have been stressed too much in the past (probably because they match both flavin and carotenoid absorption spectra reasonably well). Neither one is by any means "typical", as a first glance at Figs. 3, 14 and 23 shows, nor is the first one even a "true" action spectrum[164)]. However, all action spectra exhibit a primary sensitivity in the blue region with a second smaller peak in the near UV-region of the spectrum. For *Euglena* (Fig. 3 : 1,[48]), *Philobolus* (Fig. 3 : 2,[127]) and *Phycomyces* (Fig. 3 : 3,[40]), the null-method was used[2]. The action spectrum for the second positive curvature of the oat coleoptile (Fig. 3 : 4,[58]) was obtained by the standard method. Fig. 3 : 5,[39] constitutes the crude action spectrum for the phototropism of the avena coleoptile (plot of the phototropic response vs. wavelength at constant intensities, cf.[164]).

The defenders of the carotenoid-photoreceptor-hypothesis have always understood the shape of these action spectra in the blue to mean that the bluelight receptor is a carotenoid. Indeed, in Fig. 6 : 3 it can be observed, that the three-peak absorption spectrum of trans-β-carotenoid (in hexene) agrees well with the observed action spectrum of the avena coleoptile (Fig. 3 : 5). However, there remains one loose end which has been the crucial point of controversy in this field, ever since Galston and Baker[66] suggested in 1949 that the photoreceptor for phototropism might be a flavin: Flavin absorbs in the near UV, β-carotenoid does not.

The carotenoid hypothesis can be salvaged by the assumption, that a second unknown pigment absorbing between 350 and 400 nm (and therefore potentially fluorescing in the blue) transfers the energy via a *Förster* resonance mechanism to the carotenoid photoreceptor (such a transfer is well established in photosynthesis,[53]). Indeed, Shropshire and Withrow[162] have observed, that if coleoptiles are irradiated with light at 365 nm, blue fluorescing areas are visible near the tip with a broad emission peak near 450 nm (cf.[25]).

The basic structures of the two commonly suggested bluelight photoreceptor candidates, riboflavin and carotenoid, are shown in Fig. 4.

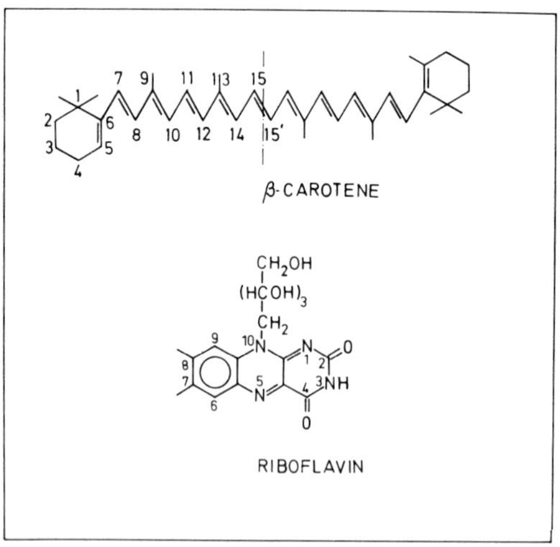

Fig. 4. The two contenders for physiological bluelight reception

2 This is essentially a compensation method analogous to electrical measurements with the *Wheatstone* bridge: The organisms are stimulated by a standard light from one side, the stimulus of which is simultaneously compensated by a second assaying light source of variable wavelength from the opposite side by adjustment of intensity.

Mostly due to technical restrictions, analysis of phototropism and other bluelight effects by action spectroscopy in the far UV-region has been quite modest. Internal screening will distort the action spectrum more seriously than in the visible and near UV part of the spectrum, because the omnipresent aromatic compounds absorb in the 250 nm region. Moreover, all UV-absorbing molecules are potential bluelight receptor candidates themselves, making UV-action spectroscopy even more doubtful. In addition, scattering depends on the inverse 4th power of the wavelength and is therefore enhanced in the UV.

Nevertheless, the avena coleoptile exhibits a curvature to unilateral UV-illumination with a satisfactory log-linear response/time relationship[38)] (the bending mode is similar to that observed for the second positive curvature which develops from the coleoptile base; cf. 2.2). Fig. 5:3[38)] shows that the double-peaked action spectrum does not match neither flavin (Fig. 5:5,[165)]) nor carotenoid absorption (Fig. 5:4,[183)]), most likely excluding both as photoreceptors. The growth hormone *auxin* (cf. 2.4 and Scheme 1) has been discussed to be a possible photoreceptor. However, in this case, this is not supported by the action spectrum either.

Phycocymes shows a positive UV-induced light-growth-response (Fig. 3:3) as well as a negative phototropic curvature, as depicted in Fig. 5:1. The exceptionally good fit of this action spectrum and auxin absorption (Fig. 5:2) might indicate that auxin is the UV-photoreceptor.

Unfortunately, no data are available comparing phototropic quantum efficiency in the far UV and the visible region. This is mainly due to the *qualitatively* different responses in both regions (tip or base curvature; 1st and 2nd positive curvature).

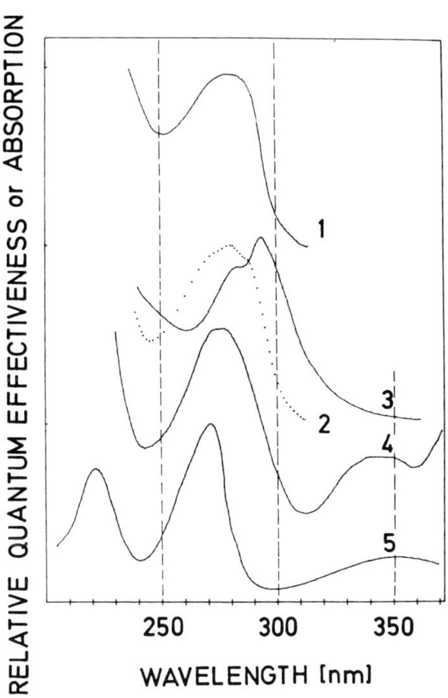

Fig. 5. Absorption and action spectra in the UV-region. Absorption spectra of: (*2*) auxin[178)], (*4*) carotene[183)], (*5*) flavin[165)]. Action spectra of (*1*) the negative phototropism of *Phycomyces*[45)], (*3*) second positive curvature of the avena coleotile[38)]

2.3.2 Modification of Absorption Spectra

Absorption spectra can be modified by several means: (1) isomeric molecules can differ in their absorption properties, (2) relatively small *intra*-molecular changes can change absorption spectra significantly, (3) *inter*-molecular interaction can induce absorption changes, (4) usually, the solvent has a great impact upon the absorption of the solute, (5) temperature influences the motion and thereby the absorption spectra of molecules, (6) binding of small molecules (coenzymes) to (apo-)proteins has a demobilizing effect, similar to low temperature, and will consequently change the absorption spectra. For each of these points, an example with respect to the flavin chromophore is discussed in this section.

It has been suggested that the photoreceptor might be the 15–15′-cis-isomer of β-carotenoid, which – in contrast to the trans-isomer – shows a UV-peak, as demonstrated in Fig. 6 : 2 and 3,[183,194]. However, its position exhibits a hypsochromic shift of 30 to 40 nm compared to known action spectra. Moreover, in the only case investigated, that of *Phycomyces*, no cis-β-carotenoid has been found[138].

Two examples among many may suffice to demonstrate the diverse spectroscopic variability of chemically modified flavin. Fig. 7 : 1 [185] shows the hypsochromic shift of 25 nm (solely) of the UV-peak of a flavin pentapeptide, isolated from monoamine oxidase by performic acid oxidation; the visible peak is not influenced. Fig. 7:2,[69] demonstrates the dramatic change of 8-acetoxy flavin upon conversion to the anion

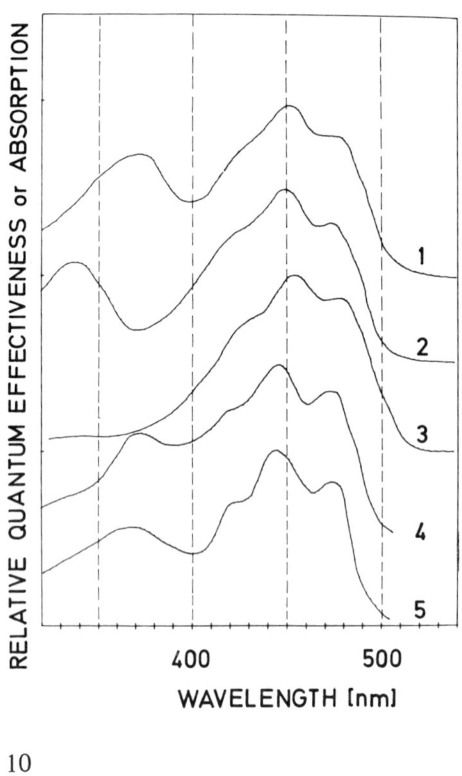

Fig. 6. Comparison of different action- and absorption spectra of (*1*) lactat oxidase[111], (*2*) β-cis carotene[183], (*3*) β-trans carotene[183], (*4*) lutein in ethanol/water[76]. (*5*): same action spectrum as in Fig. 3 : 5

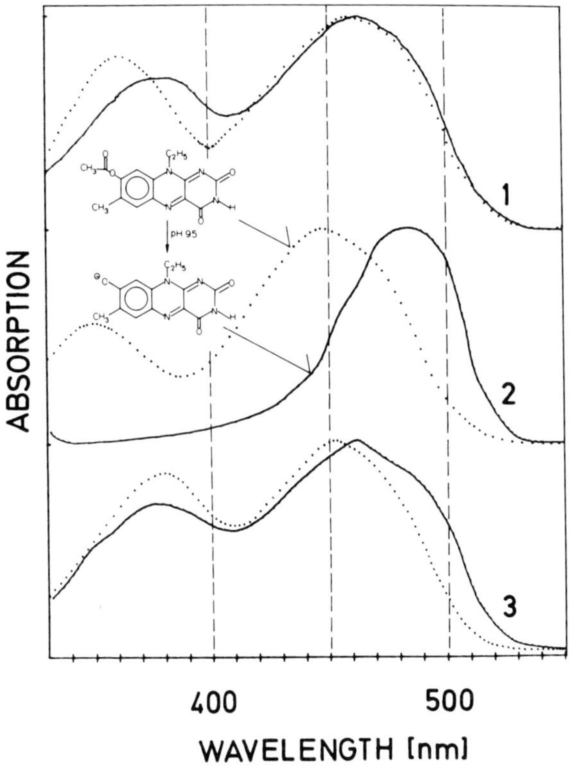

Fig. 7. Change of absorption spectra of flavins upon chemical modification. (*1*) Dotted line: pure pentapeptide from monaminoxidase; solid line: same, after performic acid oxidation[185]. (*2*) Dotted line: 8-acetoxy flavin; solid line: same, after conversion to the anion, as indicated[69]. (*3*) Dotted line: riboflavin; solid line: same, after binding to *riboflavin binding protein*[8]. For comparison, all spectra have been normalized to the same visible peak height

of the 8-hydroxy flavin: The visible peak shows a bathochromic shift of 30 nm, the peak in the near UV disappears completely.

In 1970, Hager was able to imitate the action spectrum of the avena coleoptile exceptionally well (Fig. 3:5,[39]) by adding water to an organic solution of carotenoids (Fig. 6:4[76]). However, in 1973 Song and Moore[169] demonstrated that such a water induced UV-peak is caused simply by the stacking of carotenoid molecules, bearing most likely no physiological significance.

Pigment absorption spectra are generally strongly dependent on the polarity of the solvent. This has usually been underestimated when comparing the action spectra with the absorption spectra of flavins or carotenoids, done in order to identify the photoreceptor. Bayliss and McRae[7] investigated this subject in detail and concluded that in all likelihood many factors, such as dipole-dipole, dipole-polarization, transition dipole interactions and hydrogen bonding, contribute to these effects, in an unclear manner. This, in turn, is consistent with the observation that low temperature can as well imitate the enhancement of fine structure as seen with organic solvents (Fig. 8:1,[49]). It is also consistent with the spectral change which is observed when free flavin is bound to a protein, as demonstrated in the case of *Riboflavin Binding Protein*, RBP (Fig. 7:3,[8]). Analogously, Nishimura and Takamatsu found a bathochromic shift of the visible absorption band of a carotenoid upon binding to a protein[126].

Composite action spectra characteristics of carotenoid (Fig. 8:2,[169]) and flavin (Fig. 8:1,[49]), imitated by the low temperature absorption spectra, are compared with the avena action spectrum (Fig. 8:3). Song and Moore pointed out on this basis, that the carotenoid is a rather unlikely photoreceptor, whereas the flavin is[169].

Usually, the bluelight action spectra have been compared to the absorption spectra of *free* flavins (and carotenoids) in different solvents. All UV-peaks appear at 350 nm or below, except for aqueous flavin solutions (370 nm). The latter, in turn, does not exhibit any vibrational fine-structure[77]. These features often were taken as an argument against the flavin photoreceptor hypothesis[29,40,44,162,177]. Rau was the first who compared a bluelight action spectrum (carotenoid synthesis) with the absorption spectrum of L-hydroxy-acid oxidase, suggesting a flavoprotein as photoreceptor[141]. Indeed all flavoproteins known exhibit the UV-peak between 350 and 400 nm rather than below 350 nm, due to their specific binding mode. They also have more or less pronounced vibration subpeaks in the visible range. Several flavoprotein absorption spectra are compared with the action spectrum of avena (Fig. 9). Taking the variability of action spectra and the experimental uncertainty into account, all

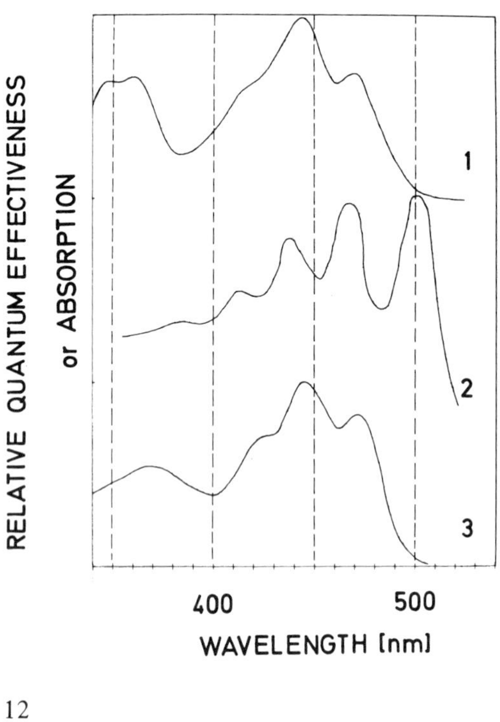

Fig. 8. "Action spectra" as mimicked by low temperature absorption spectra (liquid nitrogen) of (*1*) flavin[49] and (*2*) carotene[169]. These spectra are compared with the phototropic action spectrum (curve (*3*), same as Fig. 3:5)

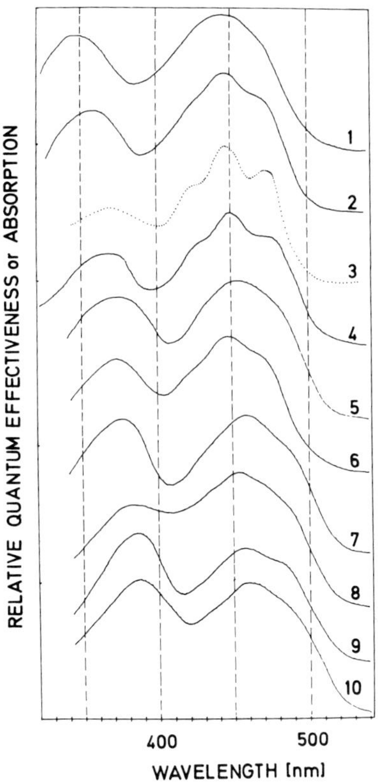

Fig. 9. Absorption spectra of several flavoproteins compared with an action spectrum of phototropism *(dotted line* (*3*),[39]). The spectra are arranged with respect to the position of their UV-peaks: (*1*) succinate dehydrogenase[78], (*2*) lipoamide dehydrogenase[179], (*4*) lactate oxidase[111], (*5*) D-amino acid oxidase[111], (*6*) flavodoxin[110], (*7*) old yellow enzyme[2], (*8*) ferredoxin $NADP^+$ reductase[16], (*9*) oxynitrilase[111], (*10*) L-amino acid oxidase[111]

flavoproteins, as e.g., lactate oxidase (Fig. 9:4) are reasonable photoreceptor candidates.

Summarizing, even a close match of any action and absorption spectra does not allow a definite identification of the photoreceptor. Additional information is required.

2.3.3 Phototropism Requires a Lateral Light Gradient

Another experimental fact on phototropism should warn us not to stress the interpretation of action spectra too much. In order to recognize the light direction, the organism primarily has to detect a light gradient (other possible mechanisms are discussed in chapter is by Doughty and Diehn of this volume). Lateral intensity gradients as small as 2% are sufficient to induce phototropism (v. Guttenberg[75]). This, however, is generated by wavelength selective extinction, as was elegantly shown by v. Guttenberg for the oat coleoptile some 20 years ago ([75], Table 1).

Therefore, the action spectrum for phototropism does not simply reflect the absorption spectrum of the active photoreceptor pigment itself, but instead, its absorption spectrum somehow modified by shading pigments. However, on this basis Thimann and Curry failed to calculate a curve fitting the experimental action spec-

Table 1. Bluelight induced bending of decapitated avena coleoptiles with different fillings, with and without application of the growth-hormone auxin (cf. Sect. 2.4). Clearly the quantity of bending depends on the intrinsic light gradient (after v. Guttenberg[75]).

Screening medium:	black paper		primary leaf		none	
indol acetic acid (auxin):	+	–	+	–	+	–
bending, degree:	28.62	8.87	25.08	5.80	11.26	1.60

trum taking a variety of concentrations of flavin and carotenoid (the main pigments) as an acceptor/shading system into consideration[177].

Using the same model, Shropshire and Withrow concluded that in the limit of *zero bending* the influence of shading pigments on the action spectrum could be completely ignored, i.e., the action spectrum is exactly equal to the absorption spectrum[162]. And only for that (extrapolated) case, the near UV-peak of the action spectrum vanishes, from which the authors conclude that the active photoreceptor is probably carotenoid in nature (cf., Fig. 6:3).

In 1966, Zenk analyzed the different pigments found in different sections of the avena coleoptile producing transverse light-gradients[195,196]. The dominating water-soluble pigments are the flavones "FO" and "FU", the latter having absorption properties identical to the presumed physiological photoreceptor "FX"[195], in the visible area, making it an excellent shadowing substance. However, taking "FX" as active photoreceptor pigment and "FU" as shadowing pigment into account, the present author failed, as well, to calculate a theoretical spectrum fitting the known action spectrum of avena (unpublished calculations).

The most photosensitive part of the coleoptile with respect to phototropism is the apical 50 μm-zone. The first millimeter of the tip is 160 times more sensitive than the second, and the second millimeter 1,800 times more sensitive than the third[108]. In 1937, Bünning[29] showed by microchemical method that there is a high concentration of carotenoid (lutein) below the tip of the avena coleoptile, from 250 μm to 2 mm; below this point and just in the extreme tip *no* carotenoid is found. This calls in question whether the enormous local differences in sensitivity can solely be caused by corresponding different absorption gradients. This doubt is further substantiated by the fact, that carotenoid deficient mutants (containing 5 to 10% of the normal siblings) show normal phototropic sensitivity.

2.4 The Role of Auxin in Phototropism of Plants

2.4.1 History

If only the extreme apex of the oat coleoptile is irradiated unilaterally, curvature nevertheless develops normally, migrating down to the light shielded region (first positive curvature). This observation led to the demonstration by Boysen-Jensen in 1910[18], showing that the *principle* causing curvature could be transmitted across a gel-filled wound gap, and was therefore chemical in nature. In 1928, Went[188] ultimately demonstrated that this chemical principle is indole acetic acid, which belongs to the general category of plant growth hormones termed *auxin* ("auxin" is now often used as a synonym for indole acetic acid). This substance is produced by the uppermost tip and transported unilaterally downwards to the lower parts which respond by elongation of the coleoptile, finally producing the curvature.

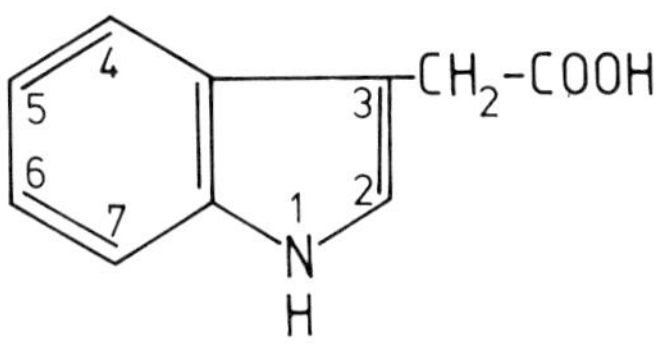

indoleacetic acid (IAA, "auxin")

Scheme 1. Molecular structure of the plant growth hormone "auxin" (indoleacetic acid, IAA). Extremely small amounts (nanomolar) can be detected by the "auxin standard test": 5 mm long segments of pea shoots elongate faster in the presence of exogenous auxin, which can be taken as a sensitive assay

Similar to phototropic bending, geotropic bending (i.e., upward bending of a *horizontally* fixed coleoptile, against gravity) is preceded by lateral transport of symmetrically produced auxin, as well. This concept probably does not hold for the geotropic response of roots, and was recently put to question[59].

2.4.2 Theories of Phototropism

Bending results from bluelight-induced one-sided enhancement of cell elongation rather than cell division. A rough estimate for the coleoptile shows that a lateral difference in growth rate as small as 1% produces about 30% curvature. From the following observations it is concluded, that bending is actually controlled by the light *distribution* within the sporangiophore *(Phycocymes)* or the coleoptile (avena), rather than the light direction itself.

First of all, when the object is irradiated simultaneously by two different bright bluelight sources which are situated at different positions, bending always occurs gradually in the direction of the resultant. Secondly, when the light beam touches the organ's surface tangentially, it bends orthogonal to the light beam, into the lighted zone (avena), respectively into the dark zone *(Phycocymes).*

Clearly, the early theories based on differential light growth response[13,14] and light-induced destruction of auxin[22,30,63,64,65,66,67,71,144,145,178] could not conclusively explain these features. Cholodny[36] and Went[188] proposed an advanced theory of phototropism invoking an *interaction* of cells as a novelty, which is widely accepted today. As mentioned above, they state that the phototropic curvature of plant organs is the consequence of a light-induced lateral/basipetal redistribution of auxin. This redistribution was elegantly demonstrated by Briggs et al.[23]. Maize coleoptile tips, placed on agar blocks, were used whole or bisected (using a thin glass barrier) as depicted in Fig. 10. The numbers given represent the amount of auxin diffused into the agar in three hours ("standard auxin test"). Clearly, the total amount of auxin eluted from the tip is independent of the illumination; its distribution, how-

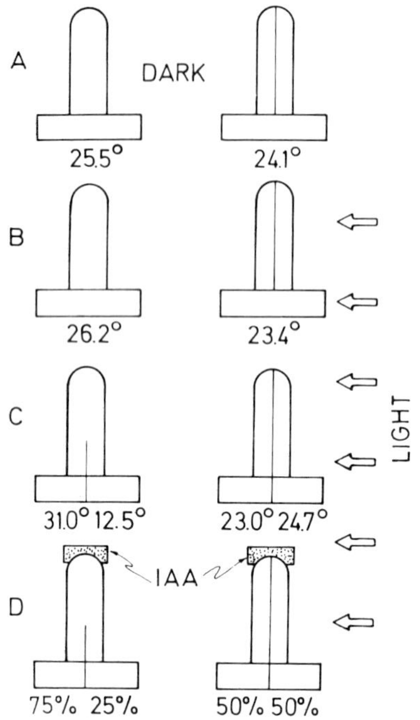

Fig. 10. Diffusion of auxin into agar blocks in three hours from corn coleoptile tips. The numbers given are curvatures obtained from a standard auxin test, i.e. a measure for the auxin diffused into the agar. (*A*) dark control: bisecting does not significantly reduce the total amount of auxin. Unilateral bluelight: (*B*) the total amount of flavin is not significantly reduced by irradiation; (*C*) partial bisecting indicates strong lateral-basipetal auxin transport[23]. (*D*) Radioactively labeled auxin, symmetrically applied to the coleoptile tip, proves the conclusion drawn from experiment (*C*)[131]

ever, is strongly light-controlled. Consistently, the lighted side is depleted while the opposite site (dark) side is enriched in auxin[188]. As expected, the same is true for the second positive curvature of the avena coleoptile, and contrary for the negative curvature[1,191]. This lateral/basipetal auxin-transport was directly proven by Pickard and Thimann[131], which applied isotopically labelled auxin to the maize coleoptile tip (Fig. 10:D). The coleoptile bisected only at the base, indicates unilateral transport by differential auxin recovery, in contrast to the completely bisected one.

A future theory on the lateral/basipetal auxin-transport has to account for the experiment by v.Guttenberg[75], who demonstrated that the presence of auxin is unnecessary for the photo-induction of a *transverse polarity* of a still unknown nature, acting as the "driving force". It therefore precedes the lateral auxin displacement. We might hypothesize that bluelight exerts some specific change in membrane permeability via a flavin moiety, thereby producing a change in membrane potential. Indeed, a light-induced current change across flavin loaded membranes has been found by Fröhlich and Diehn[62]. The nature of the transverse polarity is somewhat elucidated in the next section.

2.4.3 Electrotropism

Some thirty years ago Schrank and coworkers performed a series of electrophysiological experiments with avena coleoptiles, closely related to v.Guttenberg's experiment just mentioned. Schrank[157] succeeded to induce bending of a coleoptile by a transversal current of 10 μ amperes applied for ten minutes, which – at least initially – occurs towards the *positive* side. Depending on the distance of the electrodes from the apex, bending either levels off after a while to a constant value, or decreases again, or may even increase, as depicted in Fig. 11. Obviously, the explanation based on some kind of electrophoretic transport of the auxin anion must be rejected, because it would produce a curvature directly opposite to the one actually observed. Some alternative explanations offered by Schrank are: (1) the auxin molecule is somehow transported as a cation, (2) the applied voltage influences auxin transport by a mechanism other than electrophoresis, (3) the applied current affects the auxin-growth linkage in some way other than transport.

This latter explanation was verified by Webster and Schrank [187], who demonstrated that a transverse polarization of freshly decaptitated avena coleoptiles (containing no auxin!) could be electrically induced: *symmetrical* application of exogenous auxin to the stump *after* electrical stimulation produced curvature! A consistent behavior was exhibited if decapitated stumps, with auxin containing agar blocks on top, were electrically stimulated: bending again occurred towards the positive side, similar to the case in which auxin was applied to only one side of freshly decapitated stumps *without* electrical stimulation (Fig. 12).

That the electrically induced transverse polarization of the coleoptile is not exerted on the auxin molecule itself, is strongly supported by the fact [3], that the light-induced lateral EMF is just reversed to the externally applied one, generating the "correct" curvature: the convex side, i.e. the faster growing side, will become the *positive* side, regardless of the type of curvature (positive or negative, Fig. 13).

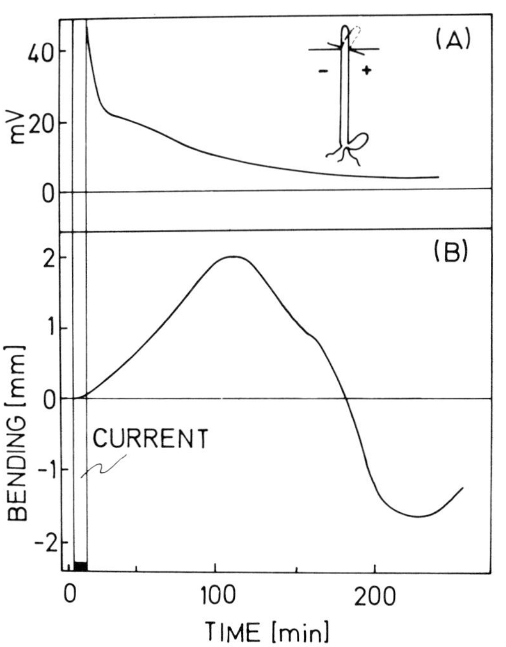

Fig. 11. Effects of transverse application of 10 μamperes for 10 min to the avena coleoptile on (*A*) the transverse electrical polarity and (*B*) the bending of the same coleoptile. Bending occurs towards the positive pole, as sketched [157]

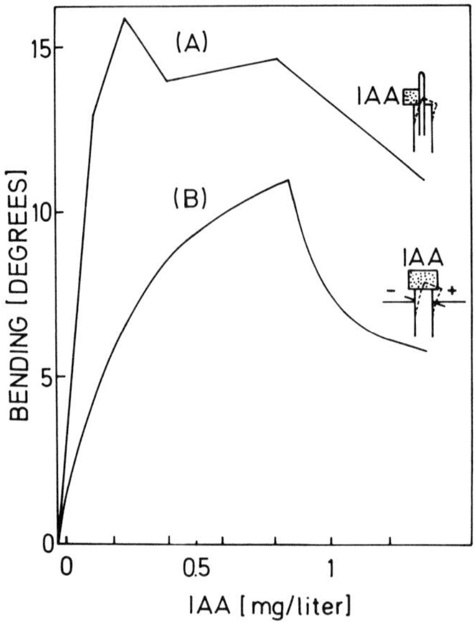

Fig. 12. Curvature responses of decapitated avena coleoptiles as a function of auxin concentration. (*A*) curvature obtained in 90 min, auxin applied asymmetrically (standard auxin test). (*B*) curvature produced in 60 min by symmetrical application of auxin upon succeeding transverse electrical stimulation. Bending occurs towards the positive pole, as sketched[187]

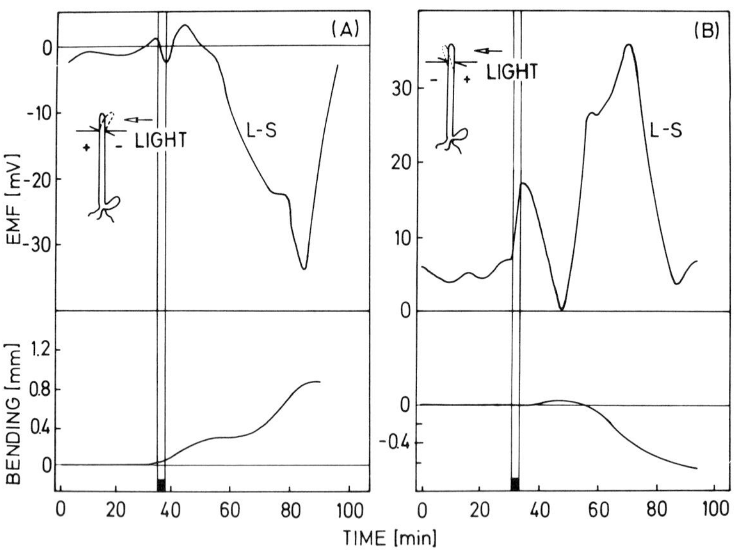

Fig. 13. Electrical and curvature responses of avena coleoptiles to unilateral irradiation of two minutes. (*A*) intensity chosen to produce positive curvature; (*B*) intensity chosen to produce negative curvature. Clearly the convex side of the coleoptile is electrically positive, regardless of the type of curvature. This indicates a strong correlation of bending and electrical potential gradient[3]

3 Bluelight Controlled Metabolism and Morphogenesis

3.1 Carotenogenesis

Those striking bluelight-induced morphological responses such as phototropism, induction of polarity of spores or rearrangement of chloroplasts, are necessarily preceded by metabolic bluelight effects. However, no clear-cut causal reaction chain has yet been elucidated.

The light dependent synthesis of carotenoids (similar to anthocyanins, cf.[43]) is probably the simplest system known, allowing molecular analysis of *primary* bluelight reactions, based on molecules capable either of suppressing or imitating the action of bluelight, or even acting as artificial light receptors. Additional information is provided by the dependence of physiological bluelight action on temperature and oxygen.

Bluelight-induced carotenoid synthesis is common in several microorganisms[6,44,91,112,141,147] and can be separated into three phases: (1) a temperature-independent light-induction, (2) a temperature-dependent period of protein synthesis which finally leads to (3) an accumulation of carotenoid pigments[78]. A few action spectra are depicted in Fig. 14 (3, *Neurospora*[44]; 4, *Mycobacterium*[91]; 5, *Fusarium*[141]). For comparison, the absorption spectrum of a crude extract of carotenoids (in petroleum ether) as induced by bluelight in *Mycobacterium* is depicted in Fig. 14:2[112]. The extract may well contain the bluelight photoreceptor itself.

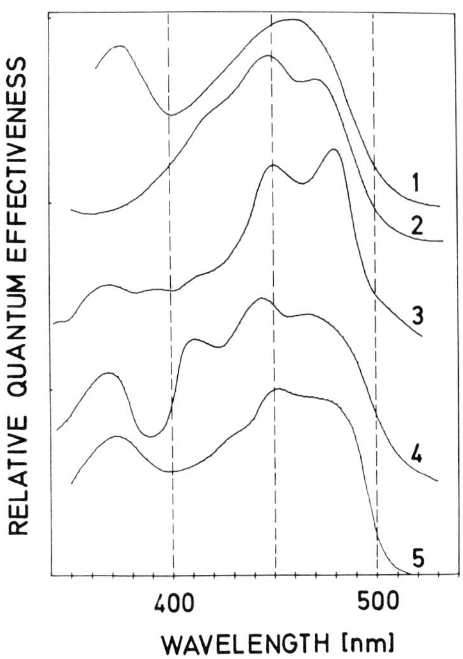

Fig. 14. Action spectra for light-induced metabolic responses. (*1*) oxygen uptake in *Chlorella*[200], and carotenogenesis of: (*3*) *Neurospora*[44], (*4*) *Mycobacterium*[91], (*5*) *Fusarium*[141]. Curve (*2*) shows the absorption spectrum of an extract of *Mycobacterium* possibly containing the bluelight photoreceptor[112]. Compare curve (*2*) to the action spectrum (*4*)

Generally, the presence of oxygen is a necessary prerequisite for the sensory transduction[91,146,193] (one exception is the oxygen-independent photoinduction of conidiation of *Trichoderma*[73]), which has been found for the phototropism of the avena coleoptile as well[75]. Therefore, it may be, that the photooxidation of some specific compound is the primary step of bluelight responses. Indeed, the time course of synthesis of various carotenoids in *Fusarium*, as induced by bluelight, could be precisely simulated by application of hydrogen peroxide[176]. In addition, Rau et al.[142,143] also found mercuribenzoate, which is known as a specific blocking agent of SH-groups to be an excellent substitute for bluelight. Consistently, the reducing agents dithionite and hydroxylamine are capable of suppressing bluelight-induction[6,176]. Therefore, it was suggested that the function of bluelight is to photooxidize SH-groups of specific proteins, such as flavoproteins[142,143]. Because oxygen is unnecessary for photochemical reactions involving carotenoid, Song et al. eliminated carotenoids from contention, favoring a flavin[168]. This redox hypothesis of the sensory transduction mechanism is strongly supported by the general observation that light-induction itself is always independent of temperature[75,79,112,136,146], (cf. 2.2).

The second step of carotenogenesis, the period of protein synthesis, has clearly been separated from the sensory transduction by means of the inhibitors of protein synthesis cycloheximide[12,78,142] and chloramphenical[12,146,147]. Regardless of their presence, light-induction was feasible, but carotenogenesis took place only *after* their removal.

The third phase of carotenoid accumulation is explored by the observation of a sequential increase of carotenoids following photo-induction. Because synthesis of sterols is *not* increased after illumination, Bindel et al.[12] concluded that the photoregulation takes place between farnesylpyrophosphate and the colored carotenoids, at least for *Fusarium.*

Another approach aiming at the mechanism of the primary photoreaction was undertaken by Lang-Feulner and Rau[107]. They incubated mycelia of *Fusarium* in photodynamically active (methylene blue, toluidine blue, neutral red) or inactive

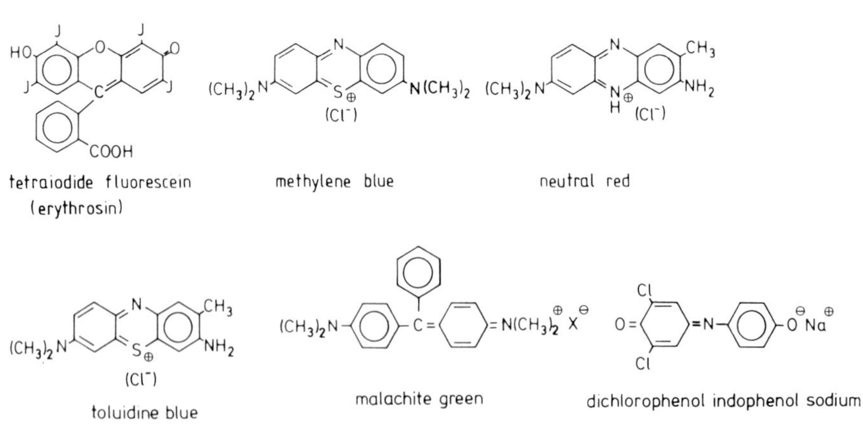

Scheme 2. Molecular structures of some photodynamically active and inactive dyes which have been employed as potential (artificial) bluelight photoreceptors

(dichlorophenol, malachite green) dyes for one hour (Scheme 2). After a 30 min irradiation with redlight, the mycelia were incubated for another two days and then tested for carotenogenesis. Clearly, only the photodynamically active dyes are capable of acting as artificial photoreceptors, supporting the hypothesis that the first step in photo-induction is an oxidative process. Similarly, Blum and Scott generated artificial, *positive* phototropism in wheat roots (which usually do not show phototropism) by introduction of the photodynamically active dye erythrosin [15]. The action spectrum clearly matched the dye's absorption spectrum. According to these authors, the dye probably acts as a photosensitizer for the oxidation of cell constituents by molecular oxygen, again consistent with the redox hypothesis of the primary photoreaction. However, a trivial photodestruction of the cells of the irradiated side (as mediated by the dye) which also would result in a "positive phototropic reaction" cannot be excluded.

3.2 Inhibitors of Photoinductive Reactions

Potassium iodide, which is known as an effective quencher of triplets and as inhibitor of excitation energy transfer [167], has been found by Mikolajczyk and Diehn [118] to be a specific inhibitor of the *inverse phobic response* of *Euglena* (a shock reaction upon decrease in bluelight intensity), as compared to the less affected rate of swimming or unaffected *direct phobic response* (a shock reaction upon increase of light intensity) and negative phototaxis (cf., Chapter by Doughty and Diehn of this volume). On the basis, the authors suggest a flavin as photoreceptor for the inverse pho-

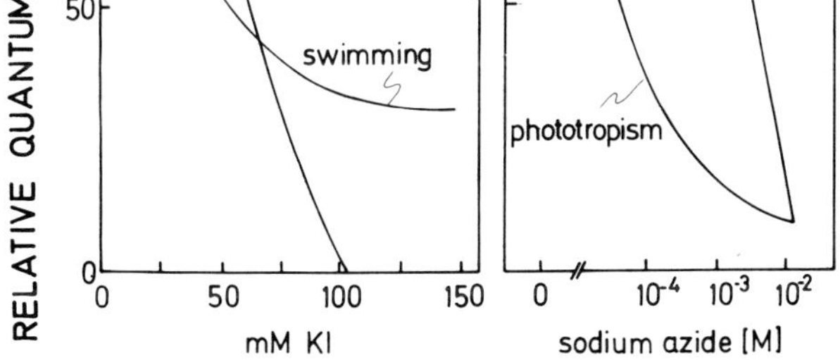

Fig. 15. Specific chemical inhibition of physiological bluelight responses: (*A*) Swimming, inverse phobic response and direct photophobic response of *Euglena* as a function of iodide concentration [118]. (*B*) Phototropic and geotropic curvature of the avena coleoptile as function of azide concentration [154]

tophobic response, but rule out a flavin with respect to the direct photophobic response, Fig. 15, A. Similarly, Schmidt et al.[154] suggested a flavin as photoreceptor for phototropism of corn coleoptiles, based on its specific inhibition by iodide, phenyl acetate (see 4.1) and azide (see below; Fig. 15B).

Batra and Rilling[6] also found up to 48% inhibition of bluelight-induced carotenogenesis in *Mycobacterium* by azide. An even higher azide effect was found by Lang-Feulner and Rau[107] in *Fusarium*, in which artificial photoreceptors had been introduced. For example, in the case of toluidine blue, carotenoid synthesis was reduced from 88 to 8% (untreated control = 100%), when azide was given to the mycelia *prior* to instead of immediately after irradiation. These results again support a flavin photoreceptor hypothesis, since azide is known for its competitive inhibition of flavoproteins (oxynitrilase[160], lactate oxidase[70], p-hydroxybenzoate hydroxylase[55]).

3.3 Photomorphogenesis of Fern

Within the life cycle of the common fern *Dryopteris filix-mas*, the young germlings originating from the haploid gonospores proceed to develop only from the transient filamentous, one dimensional stage called the "protonema", to the three dimensional stage called the "prothallus", if they receive light at a wavelength below 500 nm. On the basis of the action spectrum, Mohr[119] suggested a flavoprotein as photoreceptor for this very decisive step of differentiation. This, in turn, is most likely identical with the photoreceptor of polarotropism in the same species[171]. Most interestingly, redlight (due to phytochrome) has the same polarotropic effect as bluelight, even if it is 100 times less effective[171]. With respect to photomorphogenesis, however, redlight has just the opposite effect[119], producing even more elongated protonema than those that develop in darkness. This is probably due to different secondary reactions.

As a working hypothesis, Mohr suggested[120] that bluelight activates those genes coding for proteins which are responsible for the two-dimensional morphogenesis of fern gametophytes. This is validated by several findings. The volume of chloroplasts[51] and nuclei[9] of basal protonema cells more than doubles under blue, differentiating light, as compared to redlight[9], possibly reflecting an increase in nuclear activity[120]. Indeed, mitotic activity is strongly increased by bluelight, reflected by increasing DNA[51], RNA and protein content[50]. Consequently, incorporation of isotopically labeled carbon ($^{14}CO_2$) into proteins is doubled by bluelight, as compared to red[129, 130].

3.4 Rhythms, Flowering and Sporulation as Influenced by Bluelight

Bluelight has been found to interact with biological clocks in several modes (initiation, suppression, phase-shifting or damping of rhythms). The *Neurospora* mutant *timex* exhibits a circadian rhythm of condidiation, which is inhibited by bluelight (Fig. 16 : 2, [149]). There are two lines of evidence that prove that the photoreceptor is *not* a carotenoid favoring a flavin. First, the absorption spectrum of *timex* shows essentially a carotenoid contribution in the blue region with peak positions shifted

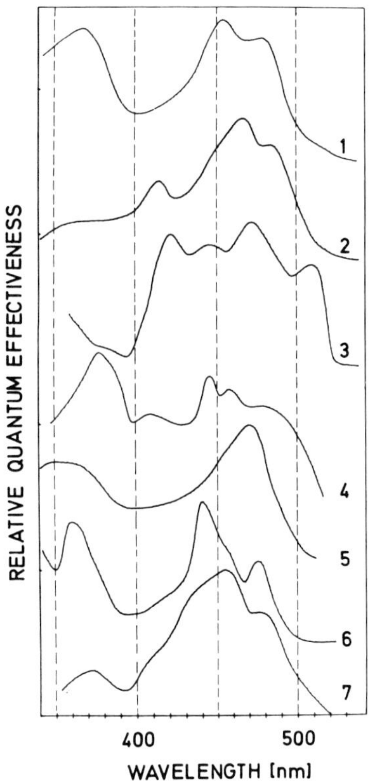

Fig. 16. Action spectra for bluelight-dependent morphogenetic responses: (*A*) Bluelight-induced chloroplast rearrangement in *Funaria* leaves[202], (*2*) bluelight-inhibited circadian rhythm of condiation of *Neurospora*[149], (*3*) bluelight-induced retardation of flower opening of *Oenothera*[148], (*4*) bluelight-induced sporolation of *Trichoderma*[72], (*5*) bluelight-induced chloroplast rearrangement of *Selaginella* cells[114], (*6*) bluelight-induced formation of perithecia in *Nectria*[41], (*7*) bluelight-induced phase delay shift of pupae emergence of *Drosophila*[99]

bathochromically by 20 nm compared to the action spectrum peaks. Secondly, the albino strain of *timex* does not reflect any carotenoid spectrally, even though the high bluelight sensitivity remains largely unchanged[149].

Bluelight-controlled responses are known for animals as well. Adult emergence in the fruitfly *Drosophila* occurs in a circadian rhythm[99,198]. A precise action spectrum for the light-induced phase shift of this rhythm was determined by Klemm and Ninnemann, Fig. 16:7,[99]. Again a flavin is favored as photoreceptor for two reasons. First, the carotenoid deficient species shows the same photosensitivity as the wild type[198]. Second, the bluelight-induced absorption change (cf., Sect. 4.2) is the same in both wild and carotenoid deficient species[99].

Another role of bluelight is exhibited by the moth *Pectinophora gossypiella*. A circadian rhythm of egg hatching can be initiated with a brief light pulse. The action spectrum (similar to Fig. 16:2,[27]) again suggests a flavin photoreceptor.

One remark is pertinent. It appears not accidental that organisms lacking a photosynthetic apparatus, such as *Neurospora, Drosophila, Pectinophora*, respond to bluelight only. On the other hand, biological clocks of species capable of photosynthesis are often controlled by light above 500 nm wavelength, indicating a dual role of photoreceptors[54,192].

A peculiar case among bluelight responses has been recognized by Saito and Yamaki, Fig. 16:3,[148]: The flower opening of the higher plant *Oenothera lamarckiana Ser.* is retarded by bluelight, however, the action spectrum clearly extends to 512 nm instead cutting off beyond 500 nm.

Two examples may serve to complete the picture of morphological bluelight responses. The asexual reproduction (sporulation) of the imperfect fungus *Trichoderma viride* and the sexual reproduction (formation of perithecia) of the fungus *Nectria haematococca* are controlled by bluelight. However, the corresponding action spectra (Fig. 16:4,[72]; Fig. 16:6,[41]) do not allow any conclusion regarding the photoreceptor. The inhibition of the light-induced formation of perithecia by *atebrin*[41, 181] does *not* hold as a pro-flavin argument either, since atebrin is known to be a relatively nonspecific enzyme inhibitor[88].

3.5 Bluelight-Induced Oxygen Uptake

Some 25 years ago, Warburg and his collaborators[186] discovered that the rate of photosynthesis of *Chlorella* under a broad band of redlight decreases after a while and the original rate is completely restored by additional, weak intensities of bluelight (less than 5% of the redlight intensity). Similar results were obtained by Terborgh[175] for the green alga *Acetabularia*. In addition, Terborgh demonstrated that the bluelight effect occurred regardless of whether blue and redlight were given simultaneously or sequentially. Therefore, the bluelight effect should not be confused with the *Emerson* dual wavelength enhancement of photosynthesis.

A bluelight-induced oxygen uptake rather than evolution, has been found in *Chlorella*[103, 104], with typical bluelight action spectra. The action spectrum, as determined by Picket and French, is depicted in Fig. 14:1,[132]. Kowallik has shown that the effect occurs for both green species and a yellow, i.e., a chlorophyll-free mutant, indicating that photosynthesis is not inhibited but that respiration is enhanced. He also demonstrated that bluelight-induced oxygen uptake does not correspond to photorespiration of higher plants, as might have been supposed. The physiological relevance for bluelight-controlled oxygen exchange remains obscure.

Taking the many action spectra for light-induced oxygen exchange into account, flavin is a better photoreceptor candidate than carotenoid. However, similar to phototropism (cf., 2.3.1), we have to consider the possibility that the two maxima in the action spectrum (at 370 and 450 nm) are based on two separate pigments. Indeed, Kowallik obtained a pigment extract of the yellow *Chlorella* mutant showing an absorption spectrum with peaks at 370 and 450 nm, and the chromatography yielded two pigments, namely, a carotenoid without any ultraviolet peak and a substance with absorption in the near ultraviolet region, but no absorption in the visible region[103].

3.6 Bluelight-Induced Chloroplast Rearrangement

Light-induced chloroplast rearrangement in several species[60,81,114,116] exhibit typical bluelight action spectra (Fig. 16:1,5). For example, bright light (> 800 erg/cm^2 s) causes chloroplasts of *Selaginella martensii*[114] to turn their edge, while weak light

(< 300 erg/cm^2 s) causes them to turn their surface towards the light source. In extremely high light intensities ($> 4 \cdot 10^5$ erg/cm^2 s) the chloroplasts contract to form clusters[114]. The physiological advantage of such behavior is obvious. From experiments with blue microbeams of 6 μm diameter (diameter of cell approx. 30 μm), Mayer[114] concluded that the photoreceptor is located in the ectoplasm.

Based on experiments with linearly polarized light, Mayer also concluded that the photoreceptor is arranged in a dichroic fashion close to the cell wall. The electrical dipole moment for the absorption of bluelight lies parallel to the cell wall, but is probably random with respect to the normal of the cell wall. In the first experiment, the cells were irradiated with bright light. Clearly, the chloroplasts separate from the walls, which are *parallel* to the *E*-vector and exhibit a banded pattern (Fig. 17, left). However, in weak polarized light the chloroplasts tended to move close to those walls parallel to the *E*-vector (Fig. 17, right).

In addition to Mayer's results, Zurzycki[199] found quite unexpectedly that polarized UV-light exerts no directing influence on chloroplasts. On the basis of the action spectrum (Fig. 16:1) for chloroplast rearrangement in *Funaria* leaves[202], he suggested a flavin as photoreceptor. Taking into account the statement of Kurtin and Song[105] regarding the flavin transition vectors, he suggested a photoreceptor-arrangement within the cell, as redrawn in Fig. 18. Such an arrangement is most likely of no advantage for the organism with respect to the photoreception process itself (input), and even less so, since natural light is essentially unpolarized. However, it might reflect some directional (cooperative?) mechanism on the *sensory transduction* side (output).

A working hypothesis concerning the primary reactions of chloroplast rearrangement has been given by Zurzycki[202]. According to this hypothesis, bluelight activates the flavoenzyme glycine oxidase, which is localized in the plasma-membrane

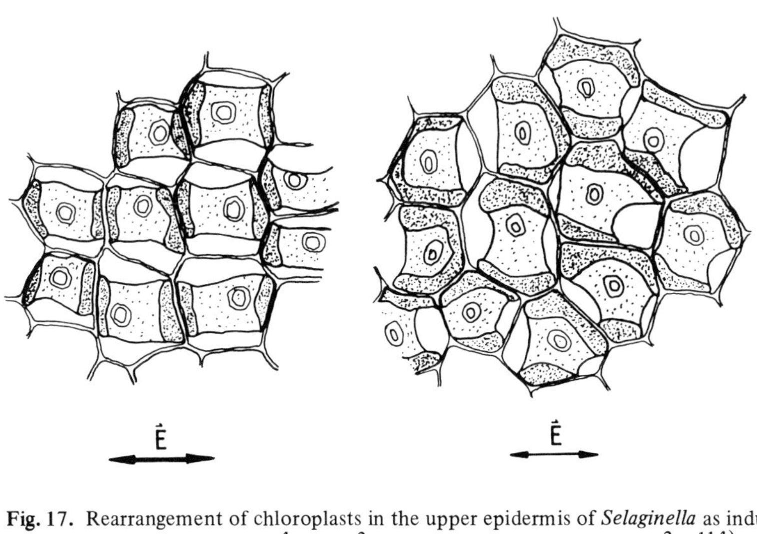

Fig. 17. Rearrangement of chloroplasts in the upper epidermis of *Selaginella* as induced by polarized: *left*, bright light ($3.5 \cdot 10^4$ erg/cm^2 s); *right*, weak light (400 erg/cm^2 s)[114]

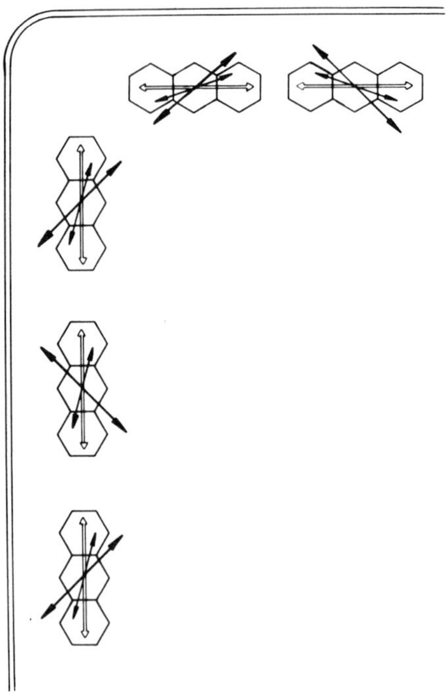

Fig. 18. Hypothetical orientation of the flavin photoreceptor within the plasmalemma in cells of *Funaria* leaves[202], based on known angles of flavin transition vectors in the blue[105]

and which is not active in darkness[150]. The chloroplast, in turn, responds *chemotactically* to the reaction products of glycin oxidation.

A dichroic embedding of the light receptor has also been found for the bluelight-induced oxygen uptake in different types of leaves[201]. *Climatium dendroides* and *Brachtythecium selebrosum* show a greater response to polarized light, when its *E*-vector is directed parallel to the leaf axis (i.e., long axis of the cells) rather than directed perpendicular to it.

Protoplasmatic viscosity, protoplasmatic streaming and rotation of chloroplasts are induced simultaneously by bluelight in the cells of *Vallisneria* leaves[161]. Similar dosage response curves and action spectra indicate that they are initiated by a single primary photochemical process. The causal sequence for the three effects is strongly suggested by the sequence of their maximum appearance after a 30 s light pulse. After 3 min, viscosity shows the lowest value, succeeded by the highest rate in streaming after 6 min, which in turn is followed by maximum phototaxis after 8 min. However, several findings demonstrate that there are at least two photoreceptors involved: (1) Dosage response curves are steeper in the blue and UV than in the red (redlight is 5 times less efficient than bluelight); (2) Polarized light is only active in the blue, not in the red; (3) The action spectrum for protoplasmatic streaming in *Vallesneria* exhibits a clear inhibition in the flavin region when treated with iodide, leaving the peak region attributed to chlorophylls a and b essentially unaffected. An additional quantity of Dichlorophenyldimethylurea (DCMU), in turn, suppresses these chlorophyll peaks, as well. DCMU and iodide are supposed to be specific inhibitors of photosynthesis and photoactivity of flavin, respectively. The main action, however, seems to be contributed by the flavin moiety[161].

In contrast to cells of water plants, cells of the avena coleoptile, as the classical object in phototropism research, exhibit an intermediate *decrease* in protoplasmic streaming upon irradiation with bluelight. The action spectrum for this response was worked out by Bottelier some 35 years ago[17]. Although this spectrum is crude, it is clear that only bluelight is active. Therefore, the effect is thought to have some bearing on the auxin transport that precedes phototropic bending (cf. 2.4),[184].

3.7 Polarotropism

In 1934, Castle discovered the phenomenon now termed polarotropism[34]. When *Phycomyces* sporangiophores are illuminated with bluelight from opposite sides, with the *E*-vector either parallel or normal to the long axis of the sporangiophore, the parallel beam must be 15% brighter than the normal one to balance the opposite tropic effects (Fig. 19). Castle himself explained the polarotropism by (trivial) different *Fresnel* reflection losses of the horizontally and vertically polarized beams, rather than by (non trivial) dichroic assembly of the photoreceptor molecules. This interpretation was experimentally confirmed by Shropshire[163]. He demonstrated that the polarization effect disappeared when reflection losses were eliminated by immersion of the sporangiophores in a liquid of identical refractivity.

This conclusion drawn by Castle and Shropshire was rejected by Jaffe[94]. He presented a theoretical analysis showing that the light lost by reflection at the surface is precisely compensated by light retained inside the sporangiophore by *internal* reflection. On the basis of several carotenoid deficient mutants, Jesaitis[96] recently

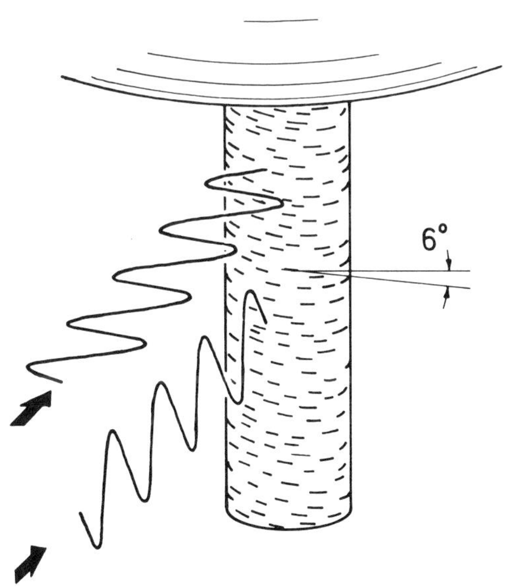

Fig. 19. Hypothetical orientation of the S0 → S1 transition dipoles (blue band) of the bluelight photoreceptor in the *Phycomyces* sporangiophore as concluded from polarized light experiments[96]. Horizontally polarized light is about 20% more effective than vertically polarized light, as sketched

confirmed Jaffe's theoretical results, in strict contradiction to Shropshire's results [163]. Using measured anisotropic action of polarized light of 486, 456 and 280 nm, Jesaitis calculated tangential orientation angles of 6° for bluelight (Fig. 19) and 33.5° for UV for the absorption dipoles. Furthermore, on the basis of fluorescence polarization data [105,166] he could show that the assumption of a *flavin* photoreceptor molecule is entirely consistent with the observed polarotropism.

In 1956 Jaffe [92] discovered a striking polarotropic bluelight response in *Fucus* zygotes (brown algae). When treated with plane polarized bluelight coming from above, they tend to germinate horizontally and in the plane of vibration. Considering both the ordinary phototropic response of zygotes to unpolarized light (the rhizoids develop opposite to the unilaterally illuminated side) and the similar dosage response curves for polarotropism and phototropism, Jaffe rejects the possible hypothesis, that the polarized irradiation orients growth by directly aligning molecules in a matrix [92]. Instead, he suggests that both effects simply are variants of the general phenomenon of phototropism. Namely, the only prerequisite is that photophysiologically active light is *non-homogeneously* absorbed within the cell (cf. 2.3.3). This non-homogeneous absorption, however, can be caused by effects such as scattering, screening, reflection and, most important and non-trivial for the bluelight receptor problem, by absorption of linearly polarized light by specifically arranged, dichroic receptor molecules. Much work has been invested in an evaluation of this latter explanation [34,93,96,114,163].

Two clear-cut, contrary examples were described by Jaffe and Etzold, exhibiting the oriented localization of the bluelight receptor in the spores of the imperfect fungus *Botrytis cinera* and the fern *Osmunda cinnamomea* [95]: *Botrytis* grows from its brighter, *Osmunda* from its darker part, when they are irradiated on only one side. On the other hand, both equally germinate in the direction of the *E*-vector of *polarized* bluelight. From such opposite behavior of physically similar spores the authors conclude that the photoreceptor molecules of *Botrytis* are strongly oriented in a dichroic orthogonal fashion, while those of *Osmunda* are parallel to the cell wall (most likely within the cell-membrane), Fig. 20.

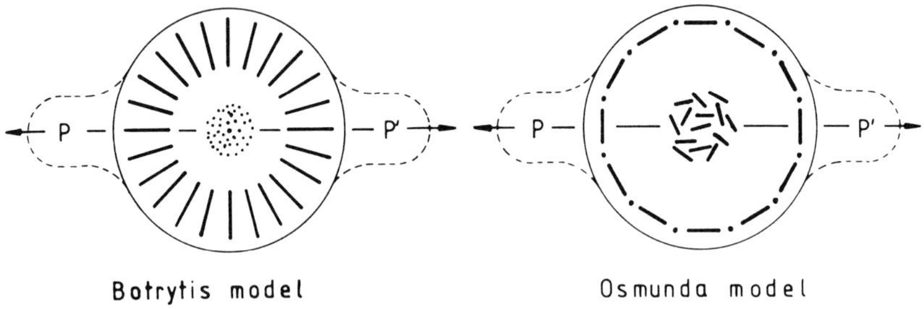

Fig. 20. Schematics of two clear-cut examples exhibiting opposite localization of bluelight photoreceptors for tropic responses, as obtained by polarized bluelight. *Left*, spore of the imperfect fungus *Botrytis cinera*; *right*, spore of the fern *Osmunda cinnamonea*. The transition dipole moments are indicated by short fat lines (dots in front view) [95]

Such an ordered organization of biological photoreceptor molecules appears to be a common feature (which, of course, is hardly detected in more opaque, highly scattering biological material). In the vertebrate rod maximum absorption lies in membranes (discs) perpendicular to the long axis of the rods, but anisotropically within these planes[47)].

The single chloroplast of the filamentous green alga *Mougeotia* is reoriented by red/far-red irradiation in an antagonistic way by means of the photoreceptor *phytochrome*, omnipresent in green plants. Haupt's results with linearly polarized microbeams clearly demonstrate dichroic orientation with the P_r-form being parallel to the cell surface and the P_{fr}-form being normal to it[83)].

The chloronema (rapidly growing germ tubes) of the fern *Dryopteris filix-mas* show similar polarotropic effects in the red and blue region, being 100 times more sensitive in the blue[171)]. Based on this observation, Etzold concluded that these effects are mediated by two different pigments, most likely phytochrome and flavin[57)]. From results with polarized light and partial illumination, he infers that both are dichroic and localized close to the cell wall with their absorption dipoles parallel to the cell surface.

4 Primary Photoreactions

4.1 Minimum Requirements for Bluelight Photoinduction

In order to be photochemically active, light must be absorbed by a specific pigment molecule. In the case for physiological bluelight-action, the most favorable photoreceptor candidate is a flavin, as discussed so far. Sun et al.[172)] assign a $\pi \rightarrow \pi^*$ character to all major flavin transitions ($S_0 \rightarrow S_1 \approx 450$ nm; $S_1 \rightsquigarrow S_0 \approx 530$ nm; $S_0 \rightarrow S_2 \approx 345$ nm; $S_0 \rightarrow T_1 \approx 595$ nm (c.f.[46)]); $T_1 \rightsquigarrow S_0 \approx 605$ nm; $T_1 \rightarrow T$ (unassigned) ≈ 650 nm[74)], c.f. Fig. 21), which are remarkably broadened by sometimes distinctive vibration modes. *Intersystem crossing* ($S_1 \rightsquigarrow T_1$) although it is for theoretical reasons forbidden, is an efficient process for flavins ($\Phi = 0.7$[74,172)]). In conjunction with relatively high fluorescence quantum yield of 25% (riboflavin/water) to 33% (riboflavin/methanol)[121)], it also accounts for the fate of the absorbed photons. However, even at 77K, phosphorescence quantum efficiency is only 0.007[121)] indicating the slowness of the radiative process and efficiency of photochemistry (the *inherent* triplet lifetime is 20 s[121)]).

It is reasonable to assume that the sensory transduction process is initiated by the flavin *triplet* state because of its relatively long lifetime of 19 μs at room temperature, compared to 2.3 ns of the singlet state[74)]. Twofold experimental evidence for this assumption was recently obtained. First, Delbrück et al.[46)] successfully induced phototropism in *Phycomyces* using extremely bright laser light (10^9 times brighter than physiologically active bluelight) between 590 and 630 nm ($S_0 \rightarrow T_1$, Fig. 22B). Even if the action spectrum deviates significantly from the flavin phosphorescence spectrum ($T_1 \rightsquigarrow S_0$),[105,172)], dotted line in Fig. 22B) this is understood by the

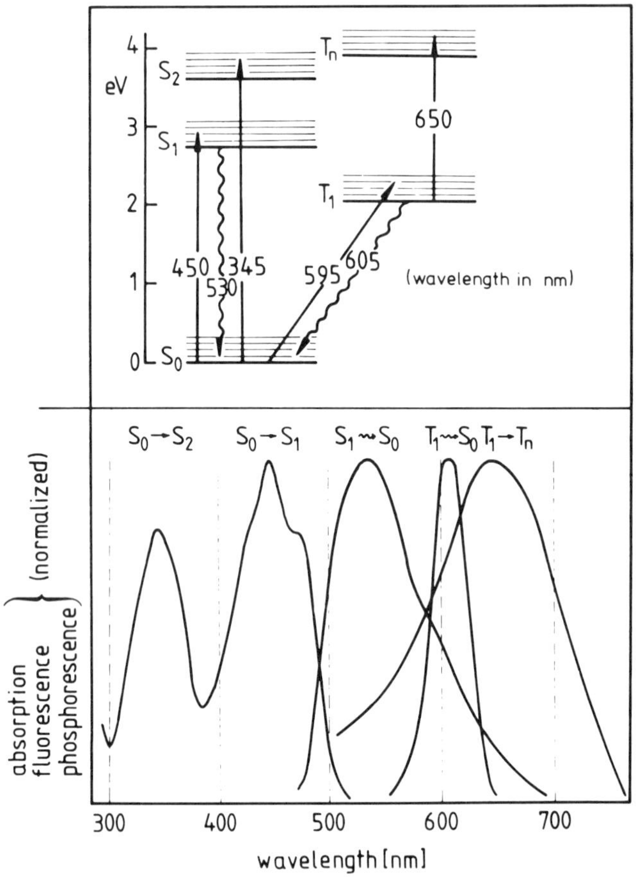

Fig. 21. *Top:* The general *Jablonski* diagram for the flavin chromophore. The given wavelengths for absorption and luminescence represent crude average values derived from the actual spectra shown below. Due to the *Franck-Condon* principle the maxima of the peak positions generally do *not* represent so-called 0 → 0 transitions, but transitions between vibrational sublevels of the different electronically excited states (drawn schematically). *Bottom:* Synopsis of spectra representing the different electronic transitions of the flavin nucleus. Differently substituted flavins show slightly modified spectra. Absorption ($S_0 \rightarrow S_2$, 345 nm; $S_0 \rightarrow S_1$, 450 nm [156]); fluorescence ($S_1 \rightsquigarrow S_0$, 530 nm [156]); phosphorescence ($T_1 \rightsquigarrow S_0$, 605 nm [105]); triplet absorption ($T_1 \rightarrow T_n$, unassigned, 650 nm [74]). For comparison, all spectra are normalized

authors as strong evidence for the involvement of the flavin triplet state in the primary photoprocess (the corresponding carotene triplet state is above 1.3 μm, [106]). The second approach was described by Schmidt et al. [154]. When maize coleoptiles are exposed to phototropically active light (449 nm, 2 μW/cm²), a simultaneous irradiation with strong, but phototropically inert, light (540–760 nm, 200 μW/cm²) will suppress phototropism up to 35% with a broad action spectrum as depicted in

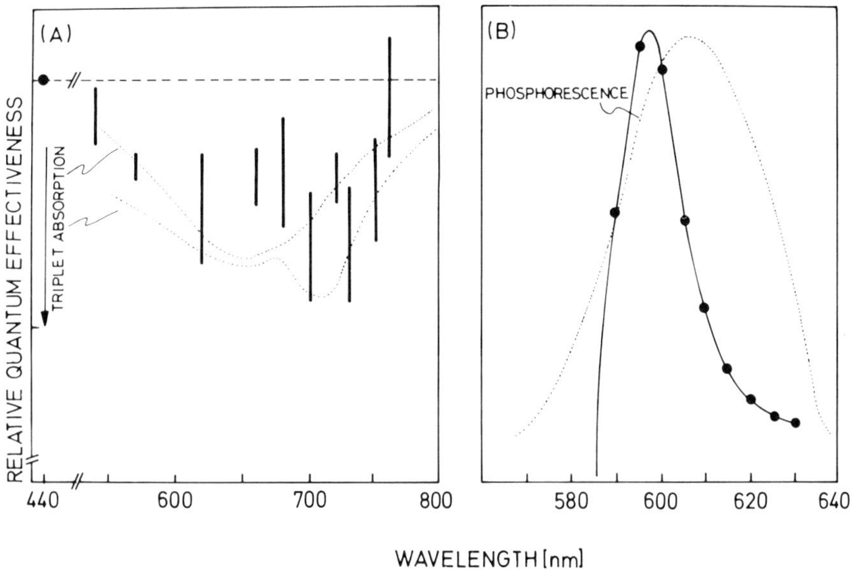

Fig. 22. (A) Comparison of flavin triplet → triplet absorption spectra (downwards drawn) with bluelight-induced (440 nm) phototropic curvature of avena coleoptiles as inhibited by strong monochromatic light in the long wave visible region [154]. **(B)** Comparison of the growth response of *Phycomyces* induced by strong laser light of wavelength longer than 590 nm [46], with the flavin phosphorescence spectrum [105]

Fig. 22A. The reasonably good fit between the measured points and the triplet-triplet ($T_1 \rightarrow T_n$) absorption spectrum of flavin (dotted lines, [174,117]) again suggests a flavin photoreceptor. These authors [154] assume an effective decrease of the lifetime of the lowest triplet state by quick triplet-triplet ($T_1 \rightleftharpoons T_n$) turnovers, thereby inhibiting photochemistry (i.e. initiation of phototropism).

For photochemical reasons carotenoids appear to be less appropriate bluelight photoreceptors. Their first excited singlet state does not fluoresce [168], indicating an extremely fast vibrational depopulation process. The lifetime of about 10^{-14} s is much too short for efficient operation of the primary photoreaction taking the rate constant of about 10^9 mol^{-1} s^{-1} for a diffusion controlled chemical reaction into account [170]. The only way to populate the triplet state is by energy transfer from a triplet donor [113], strongly suggesting that intersystem crossing ($S_1 \rightarrow T_1$) is negligible compared to internal conversion ($S_1 \rightarrow S_0$).

Nevertheless, our present knowledge does not entirely rule out a carotenoid as photoreceptor species. Carotenoids are known to work very efficiently in photosynthesis, as "light-harvesting pigments" in the blue, transferring the absorbed light energy to chlorophylls (on the basis of such "supplementary" light-receptors the available daylight-spectrum is much more efficiently utilized than with the "reaction-center-chlorophylls" alone [53,170]). On the other hand, the first excited singlet state of flavin might well be photochemically active. Because of its short lifetime, however, the prerequisite for any electron transport to occur is a *short distance* to its reaction partner. This has been provided for by Knappe[102] by an *intramolecular* photoreaction. Analogously, the natural bluelight receptor might be complexed with its reaction partner, similar to the bacteriochlorophyll (BChl) complex "P". The primary reaction of bacterial photosynthesis, the photooxidation of BChl, appears to proceed from the first *singlet* state[89].

Assuming that flavin is the actual flavin receptor, changes in quite a number of parameters might initiate sensory transduction:

1) In view of the efficient intersystem crossing, a magnetic signal is induced which might be transferred to sensitive metal centers causing spin relaxation or multiplicity changes which in turn might influence physical properties and even chemical behavior of these chromophores.
2) The change in permanent and induced dipole moments might induce an electrical signal for transfer in a given direction indicated by the change in the transition vector.
3) The rigidity of the planar flavin molecule is reduced in the excited state, which might give rise to distortions and, consequently, conformational changes of the environment.
4) The acidobasic properties of the flavin triplet are changed dramatically upon $S_0 \rightarrow S_1 \rightarrow T_1$ transition: T_1 is a base of pyridine strength (pK $\approx$ 5), while S_0 and S_1 are practically non-basic (pK $\approx$ 0),[159]. Moreover, the site of protonation is $N1/2\alpha$ for S_0 and S_1, but N5 for T_1 (Fig. 4). Hence, Hemmerich and Schmidt proposed[87] that a regiospecific rearrangement of hydrogen bridges between flavin and an apoprotein environment may occur in a photo-excited flavoprotein, which would induce a unidirectional proton transfer.
5) Trivially, photo-excitation will drastically enhance the oxidation potential of the flavin chromophore and might give rise to a great variety of reversible chemical reactions, depending on the structure of the environment and/or the pathway of potential e^-- as well as H^+-conductivity. It must be emphasized, that the oxidative action of the flavin triplet T_1 is by no means confined to $1\,e^-$-uptake from suitable aromatic π-donors such as tryptophan, tyrosin, etc.: σ-addition reactions of suitably fixed neighboring groups R—XH may occur (cf. Scheme 3):

$$Fl_{ox} + R - XH \underset{dark}{\overset{h\nu}{\rightleftharpoons}} RX - Fl_{red}H .$$

It is a wide-spread belief that such reactions could not be relevant, since they are spin-forbidden. This need not be true. It appears that the reduced flavin is a soft molecule, which resists planarity in the singlet state because of an anti-aromatic number of delocalized π-electrons. Hence, the planar conformation of Fl_{red} might have an unusually low-lying triplet state, which favors the thermal spin relaxation in $RX - Fl_{red}H$.

The main difference between e^--transfer and the last mentioned σ-addition is in the intermediates: $H\dot{F}l$ is a blue (λ_{max} = 580 nm) and planar radical, while $RX - Fl_{red}H$ is a nearly colorless ("bleached") adduct which is strongly bent out of plane in the vibrational ground state. Concomitant proton transfer might occur in both cases, if we assume that the pK of the nucleophile R – XH is essentially lower than that of RX – Fl_{red}H, which is around 6 in free aqueous solution. Combined studies searching for light-induced ESR ($g \approx 2$) and absorption (580 nm) changes in blue-light sensitive plant tissues would, therefore, be advisable. Obviously, for all these potential mechanisms, the photoreceptor pigment requires additional components to interact with, most likely a protein and/or a membrane (cf. 4.3).

4.2 Bluelight Induced Absorption Changes

Up until recently the work in the bluelight receptor field generally focussed on physiological phenomena, revealing considerable *indirect* information about the localization, orientation and nature of the photoreceptor. It is possible that the bluelight receptor today is where phytochrome was in 1959, when Butler et al.[31] successfully demonstrated the existence of *phytochrome* spectrophotometrically, which in turn initiated a dramatic development in the field of photophysiology, cf.[120]. Before the bluelight receptor problem can be attacked on a rigorous molecular basis by extraction and characterization of the pigment, a clear-cut, unique assay is required, analogous to the red/far-red photo-reversibility of phytochrome with well defined action spectra.

In 1970, such bluelight-induced spectral changes were demonstrated for the first time for *Phycomyces* by Berns and Vaughn[11] (cf., chapter Russo and Galland of this volume). However, Poff and Butler failed to reproduce these light-induced changes[135]. Instead they found another light-induced absorption change (usually abbreviated *LIAC*) that indicates an absorption increase at 430 nm and a decrease at 445 nm when freshly prepared mycelial mats of *Phycomyces* were used. However, if the sample was allowed to become anaerobic in the cuvette before irradiation, there was only a LIAC (increase) at 430 nm. LIAC's in cells of *Dictyostelium discoideum* were essentially the same[135]. Two differenct LIAC's were measured: (I) a rapidly dark decaying absorption at 410 nm, which can be induced by wavelengths as long as 600 nm (halftime 7 s[133]), and (II) a slowly dark decaying absorption at 430 nm, which is induced only by wavelengths shorter than 520 nm[135]. In contrast to LIAC II, which exhibits a pure bluelight action spectrum (similar to Fig. 23:2), only the LIAC I appears to be physiologically relevant because its action spectrum resembles very much the action spectrum of phototactic migration[133,136]. Pure bluelight effects are not known in *Dictyostelium*, suggesting that the second LIAC is physiologically meaningless.

Poff and coworkers succeeded in isolating and purifying the photoreceptor for phototaxis in *Dictyostelium*, based on action and absorption spectroscopy[134]: The pigment absorbs maximally at 430 nm and has broad absorption from 530 to 590 nm. Upon irradiation, the pigment undergoes a dark reversible LIAC (halftime of a few seconds), which in some way directs the migration of the *pseudoplasmodia*. Poff and Butler interpret this LIAC as photo-oxidation of a high spin heme pigment, which has not been further characterized[136]. The bluelight-photo-reducible pigment exhibiting the LIAC II is a b-type cytochrome similar to that in complex II of beef heart mitochondria ([137], Fig. 24:10).

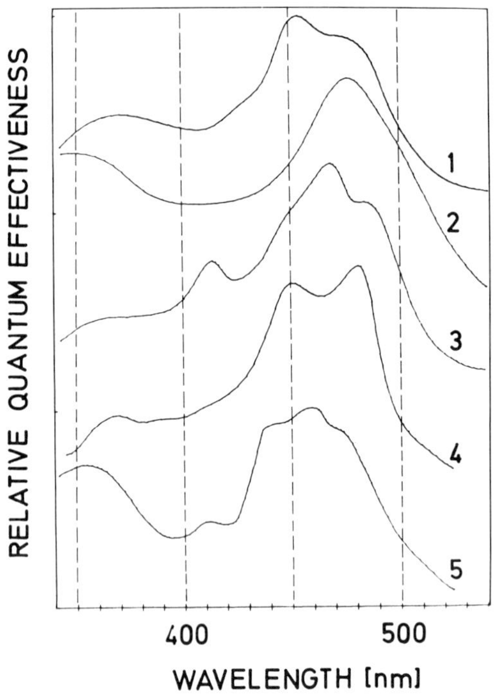

Fig. 23. Comparison of physiological action spectra with action spectra for light-induced absorbance changes in the same organisms. (*1*) Bluelight-induced light growth response of *Phycomyces*[45]; (*2*) Action spectrum for the light-induced absorbance change in the *C2*-mutant of Phycomyces[109]; (*3*) Action spectrum for the light-inhibited circadian rhythm of conidiation in the *timex*-mutant of *Neurospora*[149]; (*4*) Action spectrum for light-induced carotenogenesis in *Neurospora*[44]; (*5*) Action spectrum for the light-induced absorbance change in *Neurospora*[123]

LIAC's in *Phycomyces* were further explored by Lipson and Presti[109]. Using extremely high bluelight intensities (240 W/m^2, which is 10^6 times higher than commonly used intensities[46]), they obtained LIAC's in the cyt c/b-α-region on the order of 0.003 A for mycelial samples of the wild type, and also for several phototropically defective strains of *Phycomyces*, Fig. 24:8. The corresponding action spectrum is depicted in Fig. 23:2. The quantum yield for the cytochrome-reduction is as low as 0.015, which appears much too small to be physiologically relevant. Moreover, such a LIAC was also found in *HeLa* cells (Fig. 24:1), which are most likely not photophysiologically active. Similarly, cyt b from beef heart mitochondria has been found to be photo-reduced in the presence of high concentrations of exogenous flavin[125].

Based on such counterarguments, the authors point out that the comparability of the action spectrum for the LIAC (Fig. 23:2) and the physiological bluelight action (Fig. 23:1) is too weak of an argument to support the physiological relevance of the LIAC. Moreover, flavins are known to be extremely photo(redox)-reactive capable of reducing and -oxidizing cytochromes in vitro[152,153]. Due to the ubiquity of flavins among living organisms, an "artificial" LIAC would, therefore, not be surprising, especially when high light intensities and extremely sensitive spectrophotometrical equipment are encountered[32,35,109,138], or when the LIAC's are observed in vitro, where flavins can associate with suitable redox-partners in an unnatural manner. This leads to the discouraging statement that LIAC's are more likely the rule rather than the exceptional event[152,153]. On this basis their interpretation rather than their

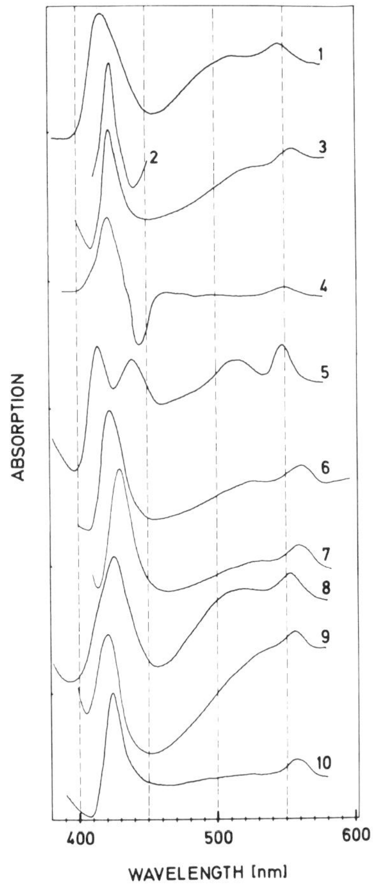

Fig. 24. Comparison of various absorbance difference spectra of cytochromes possibly involved in physiological bluelight-action. (*1*) Light-induced absorbance change in *HeLa* cells[109]; (*2*) Light-induced absorbance change in membrane-enriched fractions of corn[21]; (*3*) Low temperature oxidized reduced difference spectrum of a membrane-enriched fraction of maize[97]; (*4*) Light–dark difference spectrum for a membrane-enriched fraction of *Neurospora*[21]; (*5*) Bluelight-induced absorbance in a fraction of *Neurospora*[153]; (*6*) Bluelight-induced absorbance change of *Neurospora*, in vivo[122]; (*7*) Reduced–oxidized difference spectrum of a plasma-membrane enriched fraction of *Phycomyces*[155]; (*8*) Light-induced absorbance change of *Phycomyces*[109]; (*9*) Low temperature reduced–oxidized difference spectrum of a plasma-membrane fraction of *Dictyostelium*[155]; (*10*) Light-induced absorbance change in a cell-free extract of *Dictystelium*[135]

measurement will be the crucial point. These facts make it difficult to separate out that very flavin that is presumably functioning as a photoreceptor, as is widely assumed.

More recently, information has accumulated rendering such a separation a little more practical. Azide is known to inhibit flavin-mediated cytochrome c photoreduction as accomplished by the pathway EDTA (ethylenediaminetetraacetic acid) → flavin → cytochrome c[152]. The bluelight-induced, cyt b-photoreduction in vivo, as described by Munoz et al.[122] and Munoz and Butler[123], is also inhibited by azide[152]. On the other hand, azide is a powerful inhibitor of carotenogenesis and phototropism (cf. 3.2), supporting the idea that the primary photoaction has something to do with flavin-mediated photoreduction of cyt b. Phenylacetic acid (PAA) binds covalently in light to every flavin[85]. It has been shown, that 0.1M *exogenous* PAA can suppress 80% phototropic but only 30% geotropic bending of the maize coleoptile. Assuming that both tropisms share most of their metabolic pathway, this might reflect a specific inhibition of the flavin photoreceptor itself, offering a good chance of its labeling and isolation[154].

Klemm and Ninnemann[100] demonstrated for the first time a possible correlation between a physiological bluelight response and a LIAC (light-induced conidiation and a LIAC in *Neurospora*). Unfortunately, they observed an absorption change between 400 and 450 nm, where just no change was expected at all (cf. Fig. 24:6).

In the last two years, bluelight-induced cytochrome b-photoreduction, or at least the presence of a b-type cytochrome, has been reported for several variably purified plasma-membrane fractions. Brain et al.[21] observed such a LIAC in membrane enriched fractions from maize (Fig. 24:2) and from *Neurospora mycelium* (Fig. 24:4), with regeneration halftimes of 35 s. A light-induced photoreduction of a b-type cytochrome is *not* seen in maize coleoptiles in vivo (unpublished observation of the present author). A similar plasma-membrane enriched fraction containing a b-type cytochrome has been found by Jesaitis et al.[97] using maize. Similar to the observations by Brain et al. they found that these fractions were not mitochondrial or endoplasmatic reticular in origin. In addition, b-type cytochromes have been found in plasma-membrane enriched fractions from *Phycomyces* (Fig. 24:7), *Dictyostelium* (Fig. 24:9) and *Neurospora*[155]. A preliminary note on the photoreduction of a b-type cytochrome by a flavin in *Drosophila* larvae and pupae is given by Ninnemann and Klemm[124].

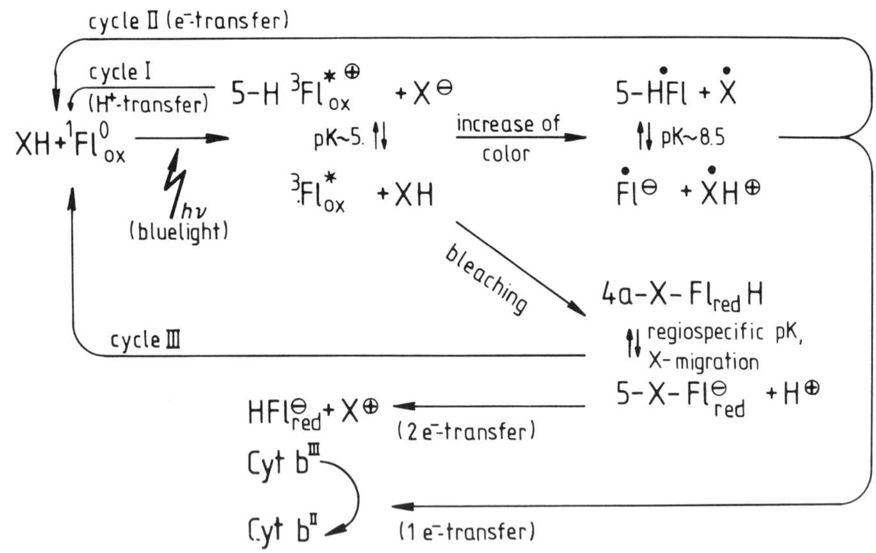

Scheme 3. Flavins are capable to undergo both 1 e⁻ *and* 2 e⁻ input/output redoxreactions as indicated. Since cytochrome b is a typical 1 e⁻ redox reagent, Hemmerich and Schmidt[86] suggest a radical mechanism of the sensory transduction (assuming that the cyt b photoreduction is phototropically relevant). The nature and fate of the flavin-photosubstrate "XH" remains obsure. For the case X ≡ cyt b the scheme represents a "reversed respiration" electron pathway

The latter authors demonstrate in their latest paper [101] that bluelight (under physiological conditions: stimulating conidiation and LIAC) will depress nitrate reductase activity and increase the activity of the smaller subunit of the enzyme complex in *Neurospora*. It is therefore suggested to be a key enzyme for physiological bluelight action.

A further promising attempt to identify the bluelight-reducible cyt b is described by Britz et al.[26] for a plasma-membrane-enriched fraction of corn coleoptiles. They find that methylene blue (Scheme 2) is capable of reducing a particular cyt b which constitutes only 10–20% of the total dithionite-reducible cytochromes. Since this particular cyt b is very similar to that which is photo-reduced by endogeneous flavin in *Neurospora* [122,123] the two are proposed to be identical.

A crude scheme of the electron pathway for the primary bluelight reaction was recently suggested by Hemmerich and Schmidt ([87], cf. 4.1):

Most likely the flavin triplet is the photochemically active species and the flavin semiquinone is the reducing agent for the b-type cytochrome.

4.3 Artificial Membrane/Flavin Systems

As discussed throughout this article, there is strong evidence that the bluelight photoreceptor is a flavin (flavoprotein) *anistropically* bound to a membrane moiety. All potential sensory transduction mechanisms require such a secondary component.

In *isotropic flavin chemistry*, i.e. the common chemistry of the free flavin nucleus in solution, the (trivial) flavin-flavin interaction ("self-contact") is always the fastest reaction [5]. This can, however, be ignored in biological systems where flavin nuclei are bound in a highly anisotropic manner to proteins and/or membranes *(anisotropic flavin chemistry)*. The flavin nucleus has four chemically active sites (positions C 4a, N5, C8 and C10a, Fig. 4), which are known to mediate practically all flavin reactions [28,86]. Regiospecificity in flavin solution chemistry depends solely upon the structure of the substrate. Under anisotropic conditions, such as in membranes, it might, however, depend primarily upon the environment. Unfortunately, very little is known about *anisotropic flavin chemistry*, which alone is applicable for the bluelight photoreceptor problem, in contrast to the isotropic flavin chemistry.

The first approach by means of a membrane model system was undertaken by Fröhlich and Diehn [62]. These authors incorporated riboflavin tetrapalmitate into lipid bilayers and found a light-induced photo-current on the order of picoamperes when an external potential on the order of 10 millivolts was applied across the membrane. The dosage response curve for this effect compares quite well with that for the photophobic response in *Euglena*. However, only extremely bright light (white, 250–4000 W/m^2) is capable of inducing the photo-response and no positive identification of the photoreceptor is given. An action spectrum is indispensable, since light-induced electrical changes in pure lipid membranes are well known: UV-light of about 300 mW/m^2 is capable of inducing photoelectric responses of 160 mV and 1.6 nA in lipid membranes made from α-tocopherol and spinggomyelin [98]. Therefore, the physiological relevance of this effect remains questionable.

A second approach with respect to anisotropic flavin (photo-)chemistry has been described by Trissl [180] and Frehland and Trissl [61]. These authors anchored flavins in artificial lipid bilayers by means of C_{18}-hydrocarbon chains at various positions of the chromophore. From fluorescence polarization analysis and model calculations they conclude, that the rotational relaxation time of the chromophore within the membrane is small compared to the fluorescence lifetime (about 2 ns [74]). They further obtain the surprising result that the chromophore is localized within the water/lipid interface, with a tilt angle of about 30° (long axis of the chromophore against the normal of the membrane), irrespective of the position where the hydrocarbon chain is bound to the flavin nucleus. They estimate an upper limit of the microviscosity of the membrane of 1 Poise.

Most recently, Schmidt [156] charged membrane vesicles with different hydrophobically substituted flavins, which yield data similar to those obtained with the flavin loaded flat membranes [61,180]. However, because of much higher total flavin concentrations methods other than fluorimetry such as absorption and electron spin resonance spectroscopy become available. This is an indispensable prerequisite for approaching the question of *sensory transduction* in bluelight-mediated physiology. Schmidt found a marked increase of flavin fluorescence at the membrane phase transition (gel → liquid crystalline), which is reversible with a significant hysteresis, depending strongly on the specific lipid/flavin system. Based on fluorescence analysis it was concluded that the rotational mobility of the flavin is strongly hindered by the lipid environment in the crystalline state, with rotational relaxation times being in the 55 to 100 ns range, depending again on the specific flavin/vesicle combination. However, beyond the phase transition the rotational relaxation times of the different flavins level off at about the same value of 10 to 20 ns, independent of the position of the aliphatic chain. The microviscosities, as seen by the chromophores, are calculated to be 120 to 200 cp below and 23 to 40 cp above phase transition. These features are taken to indicate that the chromophores sink deeper into the more hydrophobic region of the membrane upon phase transition (gel → liquid crystalline).

EPR and photochemical experiments on vesicle bound flavins seem to support these data (H. Michel and W. Schmidt; W. Schmidt and P. Hemmerich, unpublished).

5 Summary

In contrast to the other, morphologically relevant photoreceptor *phytochrome*, no unequivocal assay is available for the *bluelight photoreceptor*. This makes its purification and molecular characterization extremely difficult.

Nevertheless, there is good evidence that in all purely bluelight sensitive organisms, the photoreceptor is a flavin (flavoprotein) (Table 2), which appears to be bound to membranes (plasmalemma) in a highly dichroic manner. The mechanism of *sensory transduction* is probably correlated with light-induced redox reactions mediated by a flavin. This observation is consistent with the fact that nearly all physiolog-

Table 2. Arguments regarding the chemical nature of the bluelight photoreceptor

Pro flavin	Pro carotenoid
Action spectra usually exhibit an UV-peak between 350 and 400 nm (Figs. 3, 5, 14, 16, 23), as all flavoproteins (Fig. 9). Carotene deficient mutants generally exhibit the same bluelight sensitivity as the wild types[99,149,198]. The presence of O_2 is necessary for the *primary* photoreaction[75,91,193,146]. Light can be mimicked by oxidizing agents[142,143,176], reducing agents are capable to suppress bluelight-induction[6,176]. Photo-redox-active dyes act as artificial photoreceptors[15,107]. Light-induction is temperature independent[75,79,110,146]. Nearly all flavin reactions known are redox reactions. (Photo-)redox-reactions of carotenes are unknown; photoreductions of cytochromes are hardly mediated by carotenes. Flavin-inhibitors are capable to inhibit bluelight responses[6,107,118,154]. The low temperature absorption spectrum of isotropically dissolved flavin resembles bluelight action spectra, that of carotene does not (Fig. 8,[49,169]). Experimental evidence supports the flavin *triplet* as active species[46,154].	Albino mutants always contain minute amounts of carotene, which might serve as photoreceptor. Some action spectra show only a very small or even no UV-peak (Figs. 14:3, 16:2, 16:7). Only for the (extrapolated) limit of zero bending the action spectrum reflects the absorption spectrum of the photoreceptor for phototropism (avena coleoptile). Just for this limit the UV-peak of the action spectrum disappears[162]. The two-pigment-hypothesis explaining the two-peak action spectra (360 and 450 nm) is supported by blue fluorescing spots on the avena coleoptile, if it is excited in the UV[162]. An energy transfer from this pigment to a carotenoid is suggested. A pigment extract of the yellow mutant of *Chlorella* shows an absorption spectrum with peaks at 370 and 450 nm, which could be attributed to a carotenoid and a purely UV-absorbing pigment[103].

ically relevant flavin reactions known so far are redox reactions (unknown in carotenoid chemistry).

Several papers in the last two years attempted to separate the "photoreceptor-flavin" (and/or the associated b-type cytochrome) from the overwhelming amount of "bulk-flavin" which is involved in metabolism rather than photoreception. However, a clear-cut assay for the bluelight receptor has not yet been found. Specific photoaffinity labeling and the considerable selectivity in the photoreduction of a b-type cytochrome, as mediated by suitable dyes, are the most promising attempts.

Besides these shortcomings the bluelight receptor and sensory transduction problem is recently being attacked on the basis of completely artificial flavin/membrane systems. These appear to provide well-defined model systems to study anisotropic flavin (photo-)chemistry. This, in turn, is an essential prerequisite which allows the primary photo-events of physiological bluelight reception to be imitated and elucidated.

Acknowledgements. This work was financially supported by the Deutsche Forschungsgemeinschaft (SFB 138, A1). The author wishes to thank Drs. L.H. Pratt, V.E.A. Russo and P. Galland for reading the manuscript and for their many helpful suggestions. Prof. P. Hemmerich is thanked for his continuous interest and stimulating discussions, especially from the molecular viewpoint.

References

1. Asana, R.D.: Ann. Botany (Lond.) N.S. *2*, 955 (1938)
2. Abramovitz, A.S., Massey, V.: J. Biol. Chem. *17*, Vol. 251, 5327 (1976)
3. Backus, G.E., Schrank, A.R.: Plant Physiol. *27*, 251 (1951)
4. Ball, N.G.: In: Plant physiology, a treatise. Steward, F.C. (ed.), New York and London: Academic Press, 1969, 119
5. Barman, G.R., Tollin, G.: Biochemistry *25*, 4760 (1972)
6. Batra, P.P., Rilling, H.C.: Arch. Biochem. Biophys. *107*, 485 (1964)
7. Bayliss, N.S., McRae, E.G.: J. Phys. Chem. *58*, 1002/1006 (1954)
8. Becvar, E.B.: Ph D-Thesis, Univ. of Mich., Ann Arbor (1973)
9. Bergfeld, R.: Z. Naturforsch. *18b*, 557 (1963)
10. Bergman, K., Burke, P.V., Cerdá-Olmedo, E., David, C.N., Delbrück, M., Foster, K.W., Goodell, E.W., Heisenberg, M., Meissner, G., Zalokar, M., Dennison, D.S., Shropshire, J.R., W.: Bacteriol. Rev. *33*, 99 (1969)
11. Berns, D.S., Vaughn, J.R.: Biochem. Biophys. Res. Com. *39* (1970)
12. Bindle, E., Lang, H., Rau, W.: Planta *94*, 156 (1970)
13. Blaauw, A.H.: Rec. Trav. Botan. Neerl. *5*, 209 (1909)
14. Blaauw, A.H.: Licht und Wachstum III, Mededeelingen Landbouwhoogeschool Wageningen *15*, 89 (1918)
15. Blum, H.F., Scott, K.G.: Plant Physiol. *8*, 525 (1933)
16. Bookjans, G., San Pietro, A., Böger, P.: Biochim. Biophys. Res. Comm. *80*, 4, 759 (1978)
17. Bottelier, H.P.: Rec. Trav. Botan. neel. *31*, 474 (1934)
18. Boysen-Jensen, P.: Ber. deut. botan. Ges. *28*, 118 (1910)
19. Boysen-Jensen, P.: Growth hormones in plants. (translated by G.S. Avery and P.R. Burkholder), New York: McGraw-Hill 1936
20. Boysen-Jensen, P.: Die Elemente der Pflanzenphysiologie. Jena: G. Fischer (1939)
21. Brain, R., Freeberg, J.A., Weiss, C.V., Briggs, W.R.: Plant Physiol. *59*, 948 (1977)
22. Brauner, L.: Ann. Rev. Plant Physiol. *5*, 163 (1954)
23. Briggs, W.R., Tocher, R.D., Wilson, J.F.: Science *126*, 210 (1957)
24. Briggs, W.R.: Ann. Rev. Plant Physiol. 14, 311 (1963)
25. Briggs, W.R.: In: Photophysiology, Vol. 1, 8, Giese, A.C. (ed.), p. 223, New York and London: Academic Press 1964
26. Britz, S.J., Schott, E., Widell, S., Briggs, W.R.: Photochem. Photobiol. in press (1979)
27. Bruce, K.G., Minis, D.H.: Science *163*, 583 (1969)
28. Bruice, T.C.: Models and flavin catalysis. In: Progress in bioorganic chemistry, Vol. 4, p. 1, E.T. Kaiser, F.J. Kezdy, (eds.), New York–London–Sidney–Toronto: Wiley 1976
29. Bünning, E.: Planta *26*, 719 (1937)
30. Bünning, E., Reisener, H.J., Weygand, F., Simon, H., Kleber, J.F.: Z. Naturforsch. *11b*, 363 (1956)
31. Butler, W.L., Norris, K.H., Siegelmann, H.W., Hendricks, S.B.: Proc. Nat. Acad. Sci. *45*, 1703 (1959)
32. Butler, W.L.: Absorption spectroscopy of biological material. In: Methods of enzymology, A. San Pietro, (ed.), New York: Academic Press 1972
33. Carlile, M.J.: The photoresponse in fungi. In: Photobiology of microorganisms. P. Halladal (ed.) London: Wiley Interscience 1970
34. Castle, E.S.: J. Gen. Physiol. *17*, 751 (1934)
35. Chance, B., D. Mayer, and V. Legallis: Anal. Biochem. *42*, 494 (1970)
36. Cholodny, N.: Jahrb. Wiss. Botan. *65*, 447 (1926)
37. Chon, H.P., Briggs, W.R.: Plant Physiol. *41*, 1715 (1966)
38. Curry, G.M., Thimann, K.V., Ray, P.M.: Physiol. Plantarum *9*, 429 (1956)
39. Curry, G.M.: Ph. D. Thesis, Harward University, Cambridge, Mass. (1957)
40. Curry, G.M., Gruen, H.E.: Proc. N.A.S. *45*, 797 (1959)
41. Curtis, C.R.: Plant Physiol. *49*, 235 (1972)

42. Darwin, C., Darwin, F.: The power of movements in plants. New York: Appleton-Century 1881
43. Dawns, R.J.: Plant Physiol. *38*, 25 (1963)
44. De Fabo, E.C., Harding, R.W., Shropshire, Jr., W.: Plant Physiol. *57*, 440 (1976)
45. Delbrück, M., Shropshire, Jr. W.: Plant Physiol. *35*, 194 (1960)
46. Delbrück, M., Katzir, A., Presti, D.: Proc. Nat. Acad. Sci. *37*, 1969 (1976)
47. Denton, E.J.: Compt. rend. *239*, 1315 (1954)
48. Diehn, B.: Biochim. Biophys. Acta *177*, 136 (1969)
49. Diehn, B., Kint, B.: Physiol. Chem. Phys. *2*, 483 (1970)
50. Drumm, H., Mohr, H.: Planta *72*, 232 (1961)
51. Drumm, H., Mohr, H.: Planta *75*, 343 (1967)
52. DuBuy, H.G., Nuernbergk, E. L.: Ergeb. Biol. *9*, 358 (1932); *10*, 207 (1934); *12*, 325 (1935)
53. Duysens, L.N.M.: Nature *168*, 548 (1951)
54. Ehret, C.F.: Cold Spring Harbor Symp. Quant. Biol. *25*, 149 (1960)
55. Entsch, B., Ballou, D.P., Massey, V.: J. Biol. Chem. *251*, 2550 (1976)
56. Etzold, H.: Exp. Cell. Res. *25*, 229 (1961)
57. Etzold, H.: Planta *64*, 254 (1965)
58. Everett, M., Thimann, K.V.: Plant Physiol. *43*, 1786 (1968)
59. Firn, R.D., Digby, J.: What's New in Plant Physiology *8*, 4 (1976)
60. Fischer-Arnold, G.: Protoplasma *56*, 495 (1963)
61. Frehland, E., Trissl, H.W.: J. Membrane Biol. *21*, 147 (1975)
62. Froehlich, O., Diehn, B.: Nature *248*, 802 (1974)
63. Galston, A.W., Bonner, J., Baker, R.S.: Arch. Biochem. Biophys. *42*, 456 (1953)
64. Galston, A.W.: In: Handbuch der Pflanzenphysiologie XVII, 492–529, Ruhland, W. (ed.) Berlin: Springer 1959
65. Galston, A.W., Hand, M.E.: Am. J. Botany *36*, 85 (1949)
66. Galston, A.W., Baker, R.S.: Am. J. Botany *36*, 773 (1949)
67. Galston, A.W.: Proc. N.A.S. *35*, 10 (1949)
68. Ghisla, S., Hartmann, U., Hemmerich, P.: In: Angewandte Chemie *82*, 669 (1970)
69. Ghisla, S., Mayhew, S.G.: Eur. J. Biochem. *63*, 373 (1976)
70. Ghisla, S.: Private communication
71. Gordon, S.A., Eib, M.: Plant Physiol. *31*, Suppl. XLV, Proc. Am. Soc. Plant Physiol. (1956)
72. Gressel, J.B., Hartmann, K.M.: Planta *79*, 271 (1968)
73. Gressel, J., Bar-Lev, S., Galun, E.: Plant and Cell Physiol. *16*, 367 (1975)
74. Grodowski, M.S., Veyret, B., Weiss, K.: Photochem. Photobiol. 26, 341 (1977)
75. Guttenberg, v., H.: Planta *53*, 412 (1959)
76. Hager, A.: Planta *91*, 38 (1970)
77. Harburg, H.A., La Noue, K.F., Loach, P.A., Amick, R.M.: Proc. N.A.S. *45*, 1708 (1959)
78. Harding, R.W., Huang, P.C., Mitchell, H.K.: Arch. Biochem. Biophys. *129*, 696 (1969)
79. Haupt, W.: Planta *49*, 61 (1957)
80. Haupt, W.: „Die Phototaxis der Algen". In: Handbuch der Pflanzenphysiologie. Ruhland, W. (ed.), Berlin, Göttingen, Heidelberg: Springer 1959
81. Haupt, W., Schönfeld, J.: Ber. dtsch. bot. Ges. *75*, 14 (1962)
82. Haupt, W., Gärtner, R.: Naturwiss. *53*, 411 (1966)
83. Haupt, W.: Z. Pflanzenphysiol. *62*, 287 (1970)
84. Haupt, W.: Bewegungsphysiologie der Pflanzen. Stuttgart: Thieme 1977
85. Hemmerich, P., Massey, V., Weber, G.: Nature *213*, 728 (1967)
86. Hemmerich, P.: The present status of flavin and flavoenzyme chemistry. In: Progress in the chemistry of anorganic natural products *33*, 451 (1976)
87. Hemmerich, P., Schmidt, W.: In: Proc. of "Photoreception and sensory transduction in aneural organisms", Sept. 3–14, 1979, Versilia, Italy, in press
88. Hemker, W.C., Hülsmann, W.C.: Biochim. Biophys. Acta *44*, 175 (1960)
89. Holten, D., Windsor, M.W., Parson, W.W., Gouterman, M.: Photochem. Photobiol. *28*, 951 (1978)

90. Holmes, M.G., Smith, H.: Photochem. Photobiol. *25*, 533 (1977)
91. Howes, C.D., Batra, P.P.: Arch. Biochem. Biophys. *137*, 175 (1970)
92. Jaffe, L.: Science *123*, 1081 (1956)
93. Jaffe, L.F.: Exp. cell research *15*, 282 (1958)
94. Jaffe, L.F.: J. of Gen. Physiol. *43*, 897 (1960)
95. Jaffe, L., Etzold, H.: J. cell biol. *13*, 13 (1962)
96. Jesaitis, A.J.: J. gen. physiol. *63*, 1 (1974)
97. Jesaitis, A., Heners, P.R., Hertel, R.: Plant Physiol. *59*, 941 (1977)
98. Kay, R.E., Chan, H.: Radiation Research *40*, 177 (1969)
99. Klemm, E., Ninnemann, H.: Photochem. Photobiol. *24*, 369 (1976)
100. Klemm, E., Ninnemann, H.: Photochem. Photobiol. *28*, 227 (1978)
101. Klemm, E., Ninnemann, H.: Photochem. Photobiol. *29*, 629 (1979)
102. Knappe, W.: Electron input at the N-1 position of the flavin system. In: Flavin and flavoproteins, T.P. Singer (ed.), pp. 788–792. Amsterdam: Elsevier 1976
103. Kowallik, W.: Plant Physiol. *42*, 672 (1967)
104. Kowallik, W.: In: Photosynthesis and photorespiration. M.D. Hatch, C.B. Osmond (eds.). New York: Wiley-Interscience 1971
105. Kurtin, W.E., Song, P.-S.: Photochem. Photobiol. *7*, 263 (1968)
106. Land, E.J., Sykes, A., Truscott, T.G.: Photochem. Photobiol. *13*, 311 (1971)
107. Lang-Feulner, J., Rau, W.: Photochem. Photobiol. *21*, 179 (1975)
108. Lange, S.: Jb. wiss. Bot. *67*, 1, 1927
109. Lipson, E.D., Presti, D.: Photochem. Photobiol. *25*, 203 (1977)
110. Mayhew, S.G., Massey, V.: J. Biol. Chem. *4*, Vol. 244, 794 (1968)
111. Massey, V., Müller, E., Feldberg, R., Schuman, M., Sullivan, P.A., Howell, L.G., Mayhew, S.G., Matthews, R.G., Foust, G.P.: J. Biol. Chem. *244*, Vol. 15, 3999 (1969)
112. Mathews, M.M.: Photochem. Photobiol. *2*, 1 (1963)
113. Mathis, P., Kleo, J.: Photochem. Photobiol. *18* (1973)
114. Mayer, F.: Z. Bot. *52*, 364 (1964)
115. Mayer, F.: Zeitschrift für Botanik *52*, 346 (1965)
116. Mayer, F.: Z. Pflanzenphysiol. *55*, 65 (1966)
117. McKellar, J.F., Phillips, G.O., Checcucci, A., Lenci, F.: Excited states of flavins in photoreception processes. In: Excited states of Biological molecules. J.B. Birks, (ed.), pp. 92–105 New York: Wiley 1976
118. Mikolajczyk, E., Diehn, B.: Photochem. Photobiol. *22*, 269 (1975)
119. Mohr, H.: Planta *47*, 127 (1956)
120. Mohr, H.: Lectures on photomorphogensis. Berlin, Heidelberg, New York: Springer 1972
121. Moore, W.M., McDaniels, J.C., Hen, J.A.; Photochem. Photobiol. *25*, 505 (1977)
122. Munoz, V., Brody, S., Butler, W.L.: Biochem. Biophys. Res. Comm. *58*, 322 (1974)
123. Munoz, V., Butler, W.L.: Plant Physiol. *55*, 421 (1975)
124. Ninnemann, H., Klemm, F.: Plant Physiol. *58*, Abstr. 112 (1976)
125. Ninnemann, H., Strasser, R.J., Butler, W.L.: Photochem. Photobiol. *26*, 41 (1977)
126. Nishmura, N., Takamatsuk, K.: Nature *180*, 699 (1957)
127. Page, R.M., Curry, G.M.: Photochem. Photobiol. *5*, 31 (1966)
128. Page, R.M.: Phototropism in Fungi. In: Photophysiology Vol. 111, *3*, Giese, A.C. (ed.) New York and London: Academic Press 1968, 1965
129. Payer, H.D., Mohr, H.: Planta *86*, 286 (1969)
130. Payer, H.D., Sotriffer, U., Mohr, H.: Planta *85*, 270 (1969)
131. Pickard, B.G., Thimann, K.V.: Plant Physiol. *39*, 341 (1964)
132. Pickett, J.M., French, C.S.: Proc. N.A.S. *57*, 1587 (1967)
133. Poff, K.L., Butler, W.L., Loomis, Jr., W.F.: Proc. Nat. Acad. Sci. *70*, 813 (1973)
134. Poff, K.L., Loomis, Jr., W.F., Butler, W.L.: J. of Biol. Chem. *249*, 2164 (1974)
135. Poff, K.L., Butler, W.L.: Nature *248*, 799 (1974)
136. Poff, K.L., Butler, W.L.: Photochem. Photobiol. *20*, 241 (1974)
137. Poff, K.L., Butler, W.L.: Plant Physiol. *55*, 427 (1975)
138. Presti, D., Hsu, W.-J., Delbrück, M.: Photochem. Photobiol. *26*, 405 (1977)

139. Presti, D., Delbrück, M.: Plant Cell and Environment *1*, 81 (1978)
140. Rau, W.: Planta *59*, 123 (1962)
141. Rau, W.: Planta *72*, 14 (1967)
142. Rau, W.: Planta *74*, 263 (1967)
143. Rau, W., Feuser, B., Rau-Hund, A.: Biochem. Biophys. Acta *136*, 590 (1967)
144. Reinert, J.: Ann. Rev. Plant Physiol. *10*, 441 (1959)
145. Reisener, H.J.: Z. Botan. *46*, 474 (1958)
146. Rilling, H.C.: Biochim. Biophys. Acta *60*, 548 (1962)
147. Rilling, H.C.: Biochim. Biophys. Acta *79*, 464 (1964)
148. Saito, M., Yanaki, T.: Nature *214*, 1027 (1967)
149. Sargent, M.L., Briggs, W.R.: Plant Physiol. *42*, 1504 (1967)
150. Schmid, G.H., Schwarze, P.: Z. Physiol. Chem. *350*, 1513 (1969)
151. Schmid, G.H.: Z. Physiol. Chem. *351*, 575 (1970)
152. Schmidt, W., Butler, W.L.: Photochem. Photobiol. *24*, 71 (1976)
153. Schmidt, W., Butler, W.L.: Photochem. Photobiol. *24*, 77 (1976)
154. Schmidt, W., Hart, J., Filner, P., Poff, K.L.: Plant Physiol. *60*, 736 (1977)
155. Schmidt, W., Thomson, K., Butler, W.L.: Photochem. Photobiol. *26*, 407 (1977)
156. Schmidt, W.: J. Membrane biol., *47*, 1 (1979)
157. Schrank, A.R.: Plant Physiol. *23*, 188 (1947) (and 224)
158. Schrank, A.R.: Ann. Rev. Plant Physiol. *1*, 59 (1950)
159. Schreiner, S., Steiner, U., Kramer, H.E.A.: Photochem. Photobiol. *21*, 81 (1975)
160. Schuman-Jorns, M.: In: 6th int. symp. on flavins and flavoproteins. Kobe, Japan (1978)
161. Seitz, K.: Z. Pflanzenphysiol. *56*, 246 (1961)
162. Shropshire, Jr., W., Withrow, R.B.: Plant Physiol. *33*, 360 (1958)
163. Shropshire, Jr., W.: Science *130*, 336 (1959)
164. Shropshire, Jr., W.: Action spectroscopy. In: Phytochrome. Mitrakos, K., Shropshire, Jr., W. (eds.) London: Academic Press 1972
165. Sinsheimer, R.L.: In: Radiation biology, Vol. II. Hollaender, A. (ed.) New York: McGraw-Hill 1955
166. Siodmiak, J., Frackowiak, D.: Photochem. Photobiol. *16*, 173 (1972)
167. Song, P.S., Moore, T.A.: J. Am. Soc. *90*, 6507 (1968)
168. Song, P.S., Moore, T.A., Sun, M.: In: The Chemistry of plant pigments, 33 pp., C.O. Chichester (ed.) New York: Academic Press 1972
169. Song, P.S., Moore, T.A.: Photochem. Photobiol. *19*, 435 (1974)
170. Song, P.S., Prasad, K., Prezelin, B.B., Haxo, F.T.: Biochemistry *15*, 4422 (1976)
171. Steiner, A.M.: Naturwiss. *18*, 497–498 (1967)
172. Sun, M., Moore, T.A., Song, P.S.: J. of the Am. Chem. Soc. *94*, 1730 (1972)
173. Tauscher, L., Ghisla, S., Hemmerich, P.: Helvetica Chimica Acta *56*, 630 (1973)
174. Tegner, L., Holström, B.: Photochem. Photobiol. *5*, 223 (1966)
175. Terborgh, J.: Plant Physiol. *41*, 1401 (1966)
176. Theimer, R.R., Rau, W.: Planta *92*, 129 (1970)
177. Thimann, K.V., Curry, G.M.: Comp. Biochem. *1*, 243 (1960)
178. Thimann, K.V., Curry, G.M.: In: Light and life. McElroy, W.D., and Glass, B. (eds.) pp. 646–670, Baltimore, D.: Hopkins (1961)
179. Thorpe, C., Williams, C.H., Jr.: J. Biol. Chem. *23*, Vol. *251*, 7726 (1976)
180. Trissl, H.-W.: Biochim. Biophys. Acta *367*, 326 (1974)
181. Tschabold, E.: Phytopathology *57*, 1140 (1967)
182. Van Overbeck, J.: Botan. Rev. *5*, 655 (1939)
183. Vetter, W., Englert, G., Rigassi, N., Schwiefer, U.: In: Carotenoids. Isler, O. (ed.), p. 189, Basel und Stuttgart: Birkhäuser 1971
184. Virgin, H.I.: In: Photophysiology Vol. 1, Giese, A.C. (ed.) New York and London: Academic Press 1964
185. Walker, W.H., Kearny, E.B., Seng, R.L., Singer, T.P.: Eur. J. Biochem. *24*, 328 (1971)
186. Warburg, O., Krippahl, G., Schröder, W., Bachholzn, W., Thiel, E.: Naturforsch. *9b*, 164 (1954)

187. Webster, W.W., Schrank, A.R.: In: Archives of Biochemistry and Biophysics, Vol. *47*, p. 107, New York: Academic Press 1953
188. Went, F.W.: Rec. trav. botan. neerl. *25*, 1 (1928)
189. Went, F.W., Thimann, K.V.: Phytohormones. p. 1, New York: Macmillan, 1937
190. Went, F.W.: In: Radiation Biology, Vol. III. Hollaender, A. (ed.), p. 463, New York: McGraw-Hill 1956
191. Wilden, M.: Planta *30*, 286 (1939)
192. Wilkins, M.: Plant Physiol. *37*, 735 (1962)
193. Zalokar, M.: Arch. Biochem. Biophys. *50*, 71 (1954)
194. Zechmeister, L.: Cis-Trans isomeric carotenoids, vitamins A and arylpolyenes. Wien: Springer 1962
195. Zenk, M.H.: Z. Pflanzenphysiol. *56*, 57 (1967)
196. Zenk, M.H.: Z. Pflanzenphysiol. *56*, 122 (1967)
197. Zimmerman, B.K., Briggs, W.R.: Plant Physiol. *38*, 248 (1963)
198. Zimmerman, W.E., Goldsmith, T.H.: Science *171*, 1167 (1971)
199. Zurzycki, J.: Acta Soc. Bot. Pol. *31*, 489 (1962)
200. Zurzycki, J.: Acta Soc. Bot. Pol. *39*, 483 (1970)
201. Zurzycki, J.: Biochem. Physiol. Pflanzen *162*, 310 (1971)
202. Zurzycki, J.: Acta Protozoologica *11*, 189 (1972)

Note Added in Proof

A yearly review on "Blue Light Photoreception" covering the most recent literature, will be published by J. Gressel in the Dec. 79 issue of *Photochem. Photobiol.*

Flavins as Photoreceptor Pigments for Behavioral Responses

in Motile Microorganisms, Especially in the Flagellated Alga, *Euglena* sp.

Mike J. Doughty* and Bodo Diehn**

Department of Chemistry, University of Toledo, Toledo, OH 43606, U.S.A.

Table of Contents

Of the protozoa in which flavins have been proposed to serve as photoreceptor pigments for behavioral responses, *Euglena* has been the most intensely studied. In this contribution a survey is given of the historical studies of photomovement in this organism, followed by a critical discussion of the experimental conditions under which such studies might be carried out.

The evidence for flavin involvement as deduced from action, absorption, and fluorescence spectra, as well as via non-spectroscopic methods, is then evaluated. It is concluded that all experimental results indirectly support the flavin hypothesis, but that direct proof will have to await the isolation and in vitro characterization of the chromophore.

* *Present address:* Department of Biological Sciences, University of Cincinnati, Cincinnati, OH 45221, U.S.A.

** *Present address:* Department of Zoology, Michigan State University, East Lansing, MI 48824, U.S.A.

1 Photomovement
Observation, Analysis and Interpretation

1.1 Introduction

Alterations by light in the behavior of motile unicellular organisms (photomovement) is a phenomenon that appears to have fascinated scientists of many different callings from the outset of what we might consider modern scientific investigation. As long ago as 1817, Treviranus (cited in[7] pp. 147), reported observations on the reactions of green algae (Order *Ulotrichales*) away from light such that the cells accumulated in the shade. Concerning *Euglena*, the main subject of this review, observations date from at least 1866 on photomovement in this cell[19]. In contrast to such experimentally unsophisticated early observations on light-dependent behavior in *Euglena*, the recent literature details the application of high-technology spectral analysis to the photoreceptor of *Euglena*[10] and utilization of the Doppler characteristics of incident laser light to monitor movement of this cell[2]. Such techniques have been applied by physicists in an attempt to gain quantitative data on the movement and photomovement of *Euglena* which has consistently been the main organism of choice in studies of photomovement in unicells. However, despite over 100 years of fairly consistent attention, not too much realistic progress and understanding has been achieved in the study of movement and photomovement of *Euglena*.

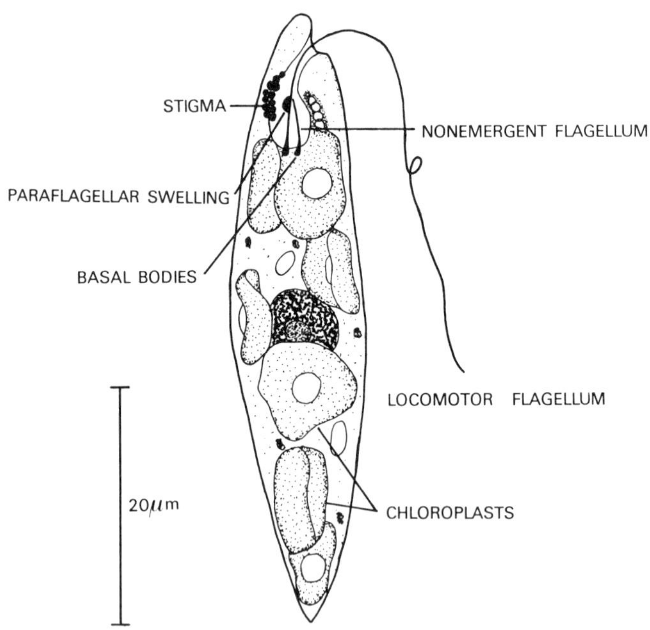

Fig. 1. *Euglena gracilis.* A diagrammatic representation of structure. After Diehn et al.[27]

Active translational motion in *Euglena* sp. (Fig. 1) is effected by either a single or several flagella[45,48,59]. Progress in the knowledge of the structure of the eukaryotic flagellar axoneme and its basic biomechanics and biochemistry has been made, especially in recent years (see monograph[85]; and recent reviews[72,92]. The flagellar axoneme of *Euglena* has the same basic characteristics as other eukaryotic flagella, i.e. the 9 + 2 microtubule structural skeleton[48,54]. It is perhaps because of such progress on other 9 + 2 axonemes that the biomechanics of flagellar motion in *Euglena* has only been the subject of very few papers[112–114]. It is rather surprising that, apart from a handful of early observations and a few preliminary investigations in recent years, all studies on photobehavior of *Euglena* sp. have been apparently carried out with little attention to the inescapable fact that light effects an alteration in flagellar activity of this cell and thus in the biochemical processes underlying this activity. It was, quite correctly, noted by Jahn and Bovee[49], that there is a serious lack of observation and systematic experimentation on the action of most physical and chemical agents on the photomovement of any photosensitive motile protozoan. The reasons for this lack of attention, as suggested by Jahn and Bovee[49] are the following. In 1838, Ehrenberg[33] considered the reddish pigmented area at the anterior end of *Euglena* sp. to be the true "eye" (the eyespot or stigma). In 1900, Wager[101] proposed that the stigma served the purpose of shading a pale, lens-like enlargement of the locomotory flagellum (the paraflagellar swelling or body, Fig. 1). Essentially, since this time, interested scientists have been preoccupied with the phenomena of "phototactic reactions" in *Euglena* (and other protozoa) and the photoreceptor mediating such behavior. It would not be unfair to conclude that such a pre-occupation has served to continually resurrect established conflicting ideas or proposals with a resultant abundance of studies designed to support or disclaim such ideas whilst revealing very little about the fundamental cause of photobehavior or its control. As contemporary reviewers, we are faced with over 100 years of sometimes supportive and sometimes inconclusive data pertaining to either the phenomena of directed movement with respect to a light source or the nature of the chromophore(s) mediating any alteration in motile behavior as effected by different incident light intensity or spectral quality. Such a plethora of information of conflicting character or conclusion, has however served to inspire no less than 15 contemporary reviews pertaining to the phenomenon of photomovement in protozoan cells. In view of this undeniable conflict of opinions and an apparent readiness on the part of recent workers to adopt an assertive standpoint with respect to both the status and knowledge of the molecular details of photomovement control in *Euglena* sp., we now feel that it would serve a useful purpose to try and constructively re-evaluate all reports on photomovement of *Euglena* sp. We hope that such a critique will help indicate why data pertaining to photomovement in *Euglena* are inconsistent and, in addition, to suggest a reasonable basis on which any future studies on either *Euglena* sp. or any other motile protozoan exibiting photobehavior, could be based. We will also show that, at this time, there are no data that would stand contemporary evaluation proving that a flavin-type chromophore mediates photomovement in *Euglena.* Nevertheless, it will be seen that available data strongly implicate a chromophore with characteristics resembling that of an isoalloxazine type moiety as being involved in either the induction or control of photomovement in *Euglena* sp. Such evidence will be critically evaluated in perspective with both the motility of flagellated cells and with photo-controlled processes in general.

1.2 Overview of Photobehavior in Unicells

A sensitivity to the blue portion of the visible spectrum is a characteristic of many cells. With some of these cells, both in vivo and in vitro spectral evidence argues against the involvement of flavins. Amongst the protozoa, these cells include:
Chlamydomonas sp.[56,74];
Volvox[64,84,83];
Gyrodinium[39,40];
Platymonas, Dunaliella and *Stephanophera*[46,47];
Cryptomonas[104,105].

Only in two microorganisms (*Dictyostelium* and *Stentor*) has *in vivo* spectral identification of the chromophore (by action spectra) been substantiated by in vitro analysis[77,78,103,110]. Since, in common with *Euglena* sp., most measurements on the spectral sensitivity of the unicells listed above, have been carried out under uncertain conditions (with respect to the biochemical state of the cells), the reader is urged to maintain an open mind prior to accepting in vivo spectral evidence as absolute. It would serve little purpose for us to add to the cycle of reviews of action spectra in motile protozoa and we refer the reader to these reviews for further information on the above listed cells (see bibliography). Flavins, largely on the basis of action spectra, have been proposed to be involved in the photomovement of *Euglena* sp., (Phylum – *Euglenophycophyta*, Order – *Euglenales*, Family – *Euglenacea*) and *Nitzschia* sp. (Phylum – *Chrysophylophyta*, Order – *Pennales*, Family – *Nitzschiaceae*). Blue-light photobehavior has been reported in procaryotes too[58,95] and potentially may involve flavins.

Since *Euglena* has consistently been the main organism in studies on photomovement, we will largely confine ourselves to studies on this cell. Before attempting to present a case for involvement of a flavin-type chromophore in *Euglena* photomovement, we will first briefly review the history of photomovement studies in this cell and review other investigations which we feel are pertinent to any such study both with respect to other behaviors of this cell and to the site and role of the chromophore in *Euglena*. Such a review will hopefully serve to highlight the problems associated with studies on *Euglena* and to provide clues as to the reasons for such diverse opinion concerning the nature of the chromophore mediating photobehavior in this cell.

1.3 Historical Survey of Photomovement in Euglena sp.

Cohn notes[19] that, upon reduction of incident white light upon *Euglena* suspensions, the cells undergo disoriented motion (ungeordnete Bewegungen). Cohn[19] also notes that *Euglena* apparently swims towards a light source and thus accumulates in a spot of light. Both Cohn and Famintzin 1867, see[52] and others[7,60] note that, if a vessel containing *Euglena* was placed such that sunlight was incident from one side only and that if the other side of the vessel was covered with a board, the cells tended to gather in a dense band on the edge of the shadow cast by the board. These two early reports qualitatively show that *Euglena* can exibit (photo) phobic (term used by Mainx[60]) reactions to changes in incident light intensity; that *Euglena* can apparently discriminate between relative intensities to thus assume a position at a favorable incident intensity and that the cells will accumulate in a spot or region of brighter light, compared to a surrounding region.

Stahl[89] notes reactions of *Euglena* sp. to changes in incident light intensity. Engelmann carefully observed the behavior of *Euglena viridis* and concluded[34] that the cells will not only accumulate in a brilliant spot of light but that they will do so as a result of a reaction to the surrounding dimly illuminated region after having initially entered the brightly illuminated region by random motion. These light spots thus acted as traps providing that the intensity of the trap was not such that the cells, by a similar shock reaction, would be prevented from ever entering the trap. Jennings[52] (pp. 136–137) also stated that *Euglena*, in a dimly illuminated region, swim about at random and can thus pass into a lighted area showing no *(apparent)* reaction on passage from the dark into the light. In the lighted region the cells also swim around at random but show an "avoiding reaction" at the light-dark boundary. By 1906 therefore, it had been established that *Euglena* shows photoaccumulation in an illuminated area, – a response mediated by the occurrence of a shock reaction that the cells undergo on trying to leave such a lighted region which they originally entered by random movement. Jennings does in fact commonly use the term "phototaxis" to describe such phenomena but, as with all other behavioral "taxes" that the reports on a variety of cells, he states quite clearly[50] that the use of such terminology was for convenience only and was not meant to designate the nature of the behavioral response underlying chemoaccumulation for example. *Euglena* has, for difficult-to-evaluate reasons, been honored with a plethora of less than self-explanatory terms to describe this common phenomenon of accumulation in a lighted region. At this time, the terminology that should be adopted, if such is deemed necessary, should probably be that of Diehn et al.[28]. The essential features of this terminology can be summarized as follows:

A response is defined as any stimulus-induced alteration in the activity of the organism's motor apparatus which can (but need not always) result in an alteration of the movement or orientation of the organism. Where nature or occurrence of a response depends upon an increase or decrease in stimulus intensity, an increase may be indicated by the use of the term "step-up" and a decrease by the term "step-down". Thus, a "step-up photophobic response" is a phobic response which occurs upon an increase in light intensity. The responses are classified as follows:

1. Kineses. These are responses in which the steady-state rate of activity of the organism is controlled by the absolute magnitude of the stimulus intensity. The sign of a kinetic response is determined relative to a null value; i.e., the activity rate in the absence of a stimulus. In a "positive" kinesis, the activity rate is greater in the presence of the stimulus than in its absence: in a "negative" kinesis, the activity rate is lower in the presence of the stimulus than in its absence. Ideally, a given kinetic response level is maintained as long as the stimulus level remains constant, i.e., adaptation does not occur.

In most kinetic responses studies to date, the linear velocity of the organism is the activity parameter that is affected by the stimulus. Since this is not the only possible type of kinesis, the following (incomplete) subdivision may be used:

a) The term "orthokinesis" indicates that the linear velocity of the organism is altered by the stimulus. In the literature on responses of microorganisms, the term "kinesis", without prefix, is often used when describing orthokinesis.
b) In "klinokinesis", the affected activity parameter is the frequency of directional change, i.e., the organism spontaneously changes its direction even in the absence of stimulation, and the stimulus modulates the frequency of such changes. "Direction" means the direction of net progress, e.g., in helically moving organisms, the direction of the helix's axis.

2. Phobic Responses. In this type of response, a change in stimulus intensity elicits a transient alteration in the activity of the organism (i.e., the response is controlled by a temporal stimulus gradient rather than by the absolute magnitude of the stimulus). The pattern of change of activity depends on the morphology of a given organism and is determined genetically and hence independent of the nature of the stimulus. The only alternative to be considered is whether in a given case the step-up or the step-down signal elicits this unique response.

A typical phobic response consists of a change in linear velocity, normally a stop response, followed by a change of direction. Unlike the situation in a kinesis, the organism resumes its original behavior after some time by means of adaptation. A typical example is the photophobic response of *Euglena*[27]. Phobic responses that have long adaptation times are nearly indistinguishable from kineses. In such cases, we suggest that the decision of whether one is dealing with one or the other be left up to the individual investigator.

3. Taxes. Taxis is a movement oriented with respect to the stimulus direction; its essential characteristic, therefore, is active orientation of the organism with respect to the stimulus source. A taxis is not a response in the strict sense of the introductory definition, but rather the *result* of a response (or of a series of responses) of the motor apparatus.

If the orientation of the organism is *toward* the stimulus source, the taxis is "positive"; if *away* from it, "negative". The term "transverse taxis" denotes *perpendicular* orientation of the organism with respect to the stimulus source.

4. Accumulation. Accumulation in or disperal from a region of higher stimulus intensity, though often confused with a true behavioral response, is the most often observed macroscopic result of many of the phenomena discussed above.

1.4 Photobehavior: Evaluation of the Action of Light on the Cell

Following the above and other early observations on the photobehavior of *Euglena* and a large number of other motile and non-motile unicellular organisms[7,52,67], attention turned to an evaluation of the spectral sensitivity of these cells (1880–1960) – a move which subsequently led to the development of exotic machinery [55] to attempt to either substantiate or disprove earlier qualitative observations on this spectral sensitivity.

We have to accept, in view of an absence of hypotheses to the contrary (apart from the very earliest literature, see [19]), that perception of incident light in any photosensitive system effects a change in the chemical activity or nature of the target site(s). A sensory signal (or signals) then effects expression of the sensory response. In other words, a chemical species in the target cell(s) or organelle(s) acts to couple the incident energy, from the visible and near-visible portions of the spectrum, to an established series of reactions already present in the cells(s) either in the active or dormant state. It is probable that these reactions occur at some predetermined level within the target cell(s) and that this activity will be both controlled by and sensitive to the previous light history of the cell. The phenomenon of artificial sensitization of many protozoa by strong chromophores has been reported in the literature to either endow photobehavior to cells not normally showing it, or to cause photodynamic damage or even premature death (see monographs [48,49,106]).

On this assumption that a light-chromophore combination, in a natural chromophore containing system, effects photosensitization of an established metabolic or physiological (locomotory) process, it is now necessary to ask in theoretical terms "what is the nature of this light-mediated process and how does incident light alter it?". Students of photobehavior have, judging from their own reports in the literature over the last 100 years, largely ignored the nature of these metabolic or physiological responses except in those cases where early experimental data supported a case for direct modulation by light of a well documented metabolic process, e.g., cyclic electron transport in the photophosphorylation systems in procaryotic cells [75].

Loeb and Mazwell [56] proposed that a given amount of stimulating energy (intensity × time) should always produce the same effect. In other words, the experimental system under investigation should exhibit a phenomenon now widely termed reciprocity. However, this will hold true only if there are no rate-limiting steps in the transduction sequence. Diehn's report [25] of reciprocity for the *step-up* photophobic response of *Euglena* over a limited range of intensities contains the only published data on this phenomenon. The experimental approach adopted with most photosensitive motile unicells has been to try and determine whether the sensory response being measured was dependent on the absolute change in incident intensity or on the rate of change of this intensity. The effects of the actual illumination (or lack of it), prior to a change in the intensity, appear to have been completely ignored. Mast and Glover report [68] that a change in the incident intensity does not effect any change in the swimming speed of *Euglena*, whilst Wolken and Shin [109] report alteration of swimming speed by light.

However, a problem of evaluation immediately arises. There has been a general tendency to describe photomovement either in terms of a positive or negative response, i.e., whether the cell's motor activity increases or decreases in response to a change in incident light intensity, or whether the cells moved towards or away from a light source. Such observation and evaluation necessarily depends on the ability of the observer or his or her equipment to detect that level of incident intensity change where a recordable response can be measured or on the self-selected definition of this "threshold" (i.e., a gross detectable response). Such "threshold" determination is largely dependent on the ability of the observer to pick out cells showing a response (i.e., his or her limits of perception) or the limit of detection by photomultiplier tubes of changes in the optical density of a cell suspension (the "phototaxigraph" principle). It should be remembered however than an ability to detect what must be regarded as a gross response is not necessarily a suitable criterion for definition of true threshold. We of course acknowledge that this is both the most widely accepted and probably the most available analytical approach. Despite this limitation however, it should be realized that the rapid response of motile cells to large changes in incident light intensity and the 'strength' of the subsequent photobehavior will not only reflect the actual perception of light, but also the kinetics of the initial chromophore-linked processes[21] and, perhaps most importantly in the case of studies on photomovement, the kinetics of the metabolic or physiological processes and their alteration by a primary hypothetical photoproduct. Many reports on photosensory behavior of unicells either have only detailed the nature of the response (positive or negative),[12,19,34,52,59,60,76,89,99]; the occurence of photoaccumulation[3,12,19,34,52,57,64,76] or the percentage of cells within a population that show a response[4,71]. At this time, we are unable to say which of these behavioral phenomena correspond to which process in the photosensory pathway, but suggest that all of them should be taken into account in order to define the photobehavior as well as possible. Of the defining parameters, the sign of the response is self-explanatory. Recording the *occurence* of photodispersal or photoaccumulation is important but, as shown in later years, the *rate* of these responses provides an indication of the relative efficiency of the stimulating light[17,23,29,55]. The percentage of cells within a screened population that shows a response probably reflects what percentage of that population have their chromophore-sensory system at a level of activation such that it can be triggered by that particular change in light intensity studied. However, such percentage sensitivity does not necessarily correspond to the overall rate of sensory photoproduct production that is surely reflected in the final sensory response, since the time dependence of stimulation, response, and recording methods is largely unknown (but see Barghigiani et al.[4]). The rate of photoaccumulation is probably determined by three factors: the percent population response, the overall motility of the cells outside the light trap and also the duration of the shock responses of individual cells. The duration of an expressed response to a given dose of incident energy perhaps most closely relates to the overall dose-response character of incident light, on the assumption that a given number of incident quanta (whether delivered as a high intensity pulse or a long duration, low intensity period of illumination) produces a given amount of a secondary photoproduct.

Feinleib and Curry[38] discuss the relationships between stimulus intensity and responses of various unicells and note that in a few unicells, the threshold perception is proportional to the initial light intensity prevailing before the change. However, although logarithmic relationships between stimulus intensity and response magnitude have been reported in flagellated algae[29,30,37], true adherence to the *Weber-Fechner Law* has not been established unequivocally because, despite this proportionality between initial intensity and the threshold perception intensity and a logarithmic relationship between stimulus intensity and response, incident doses have not always been evaluated and the responses measured have an unknown relationship to the sensory mechanism(s) (see above). Buder notes[12] that there is a quantitative relationship between stimulus intensity and response in *Euglena* (the "Resultantengesetz" – distribution of *Euglena* between two light beams of different intensity). Mast and Johnson[69] note (1932) similar relationships between the relative intensity of the two beams and the resultant trajectory of *Euglena* between them. In *Euglena*, the minimal change in irradiance (I_0 to $I_0 \pm I$) required to effect a gross change in behavior of the cell has been qualitatively observed to depend on I_0 [3,21,52,60,76,79]. Short term dark adaptation of *Euglena* has been often noted to be required for expression of strong photoaccumulation[4,17,18,23,24,44]. Any derivative experiment based upon this perception threshold, i.e., either on the intensity of light in different portions of the spectrum that can elicit a response or the intensity of light required to give a constant behavioral response in different portions of the spectrum, may thus be subject to question unless the light regime prior to the change is very carefully controlled and defined. To date, this has never been reported with *Euglena*. An additional problem facing a student of photobehavior in *Euglena* is that the induction of photobehavior can be rapidly cancelled by reversion of the light regime to the initial level or spectral quality[25]. This cancellation is observed as a fairly abrupt cessation of photobehavior (cell tumbling) and has been termed the "reset" mechanism[25]. The molecular identity of the component mediating this process remains an engima but, since resetting depends on incident dose[25], such a response indicates that the photosensory system of *Euglena* possesses characteristics uncommon to other photoreceptive systems. It is interesting to note that Mast proposes[66] that there are two types of light response in *Euglena:* one of which determines photoaccumulation and depends on the rate of light intensity change (see also Creutz[21]); the other depends on the energy levels received and determines the degree of activity (strength) (see also[56]). Experimental clarification of these points remains to be accomplished.

1.5 Photomovement: Is there Phototaxis in Euglena?

Before evaluation of the behavior of *Euglena* further, it is necessary to pose the question as to what level of complexity exists in *Euglena* for control of light-dependent behavior. Alteration in the swimming behavior ("disoriented motion", "shock responses", "shock reactions", avoiding reactions", "illumination reactions", "phobic reactions") following sudden alterations in either incident sunlight or white light have been repeatedly reported[3,19,34,51,52,60,63,99]. However, apart from descriptions provided by Cohn[19], Engelmann[34], Oltmans[76] and Bancroft[3], a lengthy discussion of the nature of such behavioral responses by Mast[64] and some rather elegant diagrams by Jennings[52], there have been few quantitative descritions of the cell behavioral sequences executed by *Euglena* under controlled conditions. Descriptions of the initial characteristics of the shock responses are available from microscope observations on either cell or flagella activity[3,19,34,50,51,61,63,89,99] or from high magnification cine film of flagellar activity of single cells[27].

In addition however, several authors[3,19,52,64], provide us with detailed descriptions of apparent oriented movement of *Euglena* towards a light source. Jennings describes[52] a series of behavioral responses (course corrections in swimming path) that apparently permit *Euglena* to readopt a swimming trajectory towards a light source after displacement of this source through an angle of 180°. Several authors[12,63,69] report that *Euglena* adopts a diagonal trajectory between two beams of light set at 90° to each other. Buder[12] and Mast and Johnson[69] also note that the trajectory depended on the relative intensity of the two beams. Actually, in *Euglena gracilis* we have seen oriented movement of the type described by earlier workers (i.e., *positive* taxis toward light of moderate intensity) only very rarely. This organism will, however, exhibit distinct *negative* phototaxis i.e. oriented swimming away from the light, when exposed to light of sufficiently high intensity (1 kW/m^2 or more). We believe that positive phototaxis as a common phenomenon, if it exists at all, is restricted to *Euglena* species or environmental conditions that we have not studied.

1.6 Positive Versus Negative Photobehavior in Euglena

Early workers note the existence of a "negative" reaction of *Euglena* to incident light, i.e. repulsion from light or shock reactions upon illumination rather than reduction in light intensity. (But note that in the presently accepted terminology, this is termed a "step-up" rather than a "negative" response). Jennings for example ([52], pp. 140) notes that, if *Euglena* "are suddenly subjected to sunlight, they give the avoiding reaction". No other details are given. He also notes however ([52], pp. 141) that "direct sunlight is very injurious to it (the cell); if long continued it causes the organism to fall to the bottom and die". Bancroft[3], during studies on such "shock reactions" occuring on sudden illumination of the cells, notes (pp. 408–409) that no shock reaction could be produced by illumination periods of 0.1 second, (in contrast, Diehn[23,24] finds that a population will respond to a millisecond flash). Bancroft notes a delay between changes in incident light intensity and the "illumination reaction" of 0.4–0.5 s maximum, but more than 0.1 s. He concludes, "Consequently, since effective stimulation must take place within the time interval of 0.3 s, it cannot be the shock movement mechanism that is stimulated"... "the time required for effective stimulation evidently varies inversely with the light intensity ..." ([3], pp. 408). Such observations and conclusions suggest that light does not simply act as a switch to alter flagellar activity; neither does light directly alter the flagellar activity.

Since *Euglena* sp. are known to exhibit reactions to both increases in incident light intensity ("shock reactions", "avoiding reactions", "illumination reactions") and to decreases in incident light intensity ("disoriented motion", "shock reactions", "Photophobic response")[3,19,34,51,52,61,63,89,99], it is perhaps not surprising to find reports of the occurence of no apparent response following some light intensity changes. This is usually observed as a lack of reaction of a population of *Euglena* to variations in incident light intensity or to a light trap boundary. Bancroft ([3], pp. 399–400), notes the existance of a narrow "indifference zone" i.e., an intensity range of incident light to which *Euglena* shows neither shock reactions nor "heliotropism" (photoaccumulation), but only under some culture conditions. Buder[12] too notes such an "Indifferenzzone" in which *Euglena* are randomly oriented with respect to light. Diehn[25,26] interprets this indifference zone in terms of a double threshold and as being indicative of an identifiable molecular component in his hypothesised sensory transduction scheme. However, it must be stressed that such zones are not fixed with respect to incident irradiance and have been defined as being zones where no *gross* responses of the cells are observed. Bancroft notes[3] that the "illumination reaction" can be fatigued by illumination of the cells with intense light for 30–45 minutes see also[98]. Such a phenomenon serves to demonstrate that the "indifference zone" is "floating" and very much depends on the previous light history of the cell as well as probably on a multitude of other factors such as cell growth stages, environmental conditions etc. As discussed above, such indifference zones in both photoaccumulation and photodispersal will probably alter any action spectra because such an indifference zone will be closely related to the determination or detection of any net population response.

1.7 Effect of Extracellular Environment on Photobehavior

Even a casual review of the solutions in which photobehavior of *Euglena* (or any other unicell) have been studied, serves to provide the contemporary analyst with both entertaining reading (descriptions of a garden pond for example: Wager[102]) and, in addition, a perplexing view of photobehavior protocol. Since the initiation of systematic sensory physiology studies in the last century, studies on sensory responses (in this case, of muscle) were carried out in selected and defined extracellular bathing solutions. The classic Ringer Solution for muscle physiology and electrophysiology was developed in the early years of this century. Sensory behavior in other protozoa has been largely studied in relatively defined extracellular medium although we must acknowledge that such attention to medium conditions was largely the result of the fact that chemosensory behavior was being studied (see review by Dryl[32]). However, we can but ask at this time, especially in view of the continual presence of notes on the effects of different chemicals and conditions on photobehavior, why no determined attempt has been made to evaluate the photobehavior of *Euglena* under even relatively defined extracellular conditions (other than Diehn's "resting medium"[23], which was employed to stabilize the cell's photoresponses). With respect to specific chemical effects, Massart[61], during some lengthy studies on the sensitivity of many eukaryotic and procaryotic unicells to salt solutions, notes the dramatic alteration of photobehavior of *Euglena* (movement towards light) upon swimming close to a drop of concentrated salt solution. Mast notes the effects of a large variety of chemicals on the photobehavior of a colonial alga[65]. Frank[42] and Mainx[59,60], as part of studies on a variety of photosensitive protozoa, note dramatic alterations in the behavior of *Euglena* to mineral and organic acids, in addition to a variety of other chemicals. Mainx also notes[59] the effects of many chemicals (acids, anesthetics, aliphatic alcohols and inorganic salts) on the behavior of several photosensitive protozoa. Frank investigated chemosensory behavior[42] in *Euglena* (and *Chlamydomonas*) by observing their reaction to capillary tubes filled with salts, acids and amino acids, by studying whether they move into the tubes and whether the response was strong or weak. Mainx details and cites a considerable number of observations on chemosensory behavior of *Euglena*[60]. Such responses were either shock reactions or accumulation round the mouth of capillary tubes containing chemicals (acids, salts, alcohols).

Whilst such reports only provide suggestive indications that chemosensory behavior to, for example, salts or amino acids, may directly modulate photobehavior of *Euglena* (and it should be remembered that most contempory studies on photobehavior of *Euglena* have been carried out in a culture fluid rich in amino acids and salts), another chemosensory response of *Euglena* is well documented to dramatically alter both the expression of photosensory behavior (in particular its sign) and to effect an apparent alteration in the spectral sensitivity of *Euglena*. This phenomenon has been loosely termed "aerotaxis" and has been observed (as accumulation of many protozoa around an air bubble in cell suspensions) by many workers[41,50,53]. Sensory behavior to dissolved gasses in suspending medium, usually following conditions serving to deplete the oxygen levels in the medium, has been documented in *Euglena*[1,20,60,79,102] (in agreement with the latter, Jennings and Moore report[53] that *Euglena viridis* does not show responses to a bubble of CO_2). Such "aerotaxis" has been reported to alter the sign of phototaxis in *Euglena*[12,60,79,102]. This chemosensory behavior is probably responsible for the apparent photoaccumulation of *Euglena* into a red light trap[16,22,25,108].

Strassburger (1878; cited in[7,48,49]) reported that any rise in the temperature of a suspension of various photosensitive cells, without change in illumination intensity, tended to produce positive phototaxis and *vice versa*. Mast[63,64] notes that *Euglena* showing strong photoaccumulation in a beam of constant light intensity, became less active as the temperature decreased, came to rest at 10 °C and became more active at lower temperatures. Diehn and Tollin[29] report alterations in the gross rate of accumulation of *Euglena* in a monochromatic light trap as a function of temperature.

In recent years (1957–1979), there have been several reports of more specific alterations of photobehavior in motile alga by cations. Halldal, in a series of studies from 1957–1962 on the photosensory behavior of several algae[46,47], notes that the sign of phototaxis and the relative magnitude of phototaxis could be altered by Mg^{2+} ions and Ca^{2+} ions and more especially by the relative concentrations of these two ions in the extracellular medium. Stavis and Hirschberg[90] made a preliminary study of the effects of various inorganic cation salts on the motility and "phototaxis" of *Chlamydomonas*. Their data show that low concentrations of inorganic cation salts or their combination are required for maximal expression of photosensory behavior. The effect of inorganic cation salts on the behavior or photobehavior of *Euglena* apparently has not been considered seriously, as evidenced by a lack of attention paid to such effects[42,59,60]. Such reports, on alteration on behavior and photobehavior of protozoa by inorganic cations, although generally qualitative in nature, do however indicate that such cations can alter the photosensory/sensory transduction mechanisms in these cells. Such alterations as arise as a result in differences in the extracellular media in which photobehavior has been studied, could, in part, explain why photoaccumulation characteristics and the spectral sensitivity of *Euglena* appears to be different in different reports. An illuminating example of such interactions between inorganic ions and photosensory systems is provided by the ciliated protozoan, *Sten-*

tor, which shows both marked photosensory behavior[52,62] and chemosensory behavior to inorganic cation salts[70,76a,94,111].

2 The Chromophore: Analysis, Localization and Sensitivity

2.1 Spectral Sensitivity of Euglena sp. and Early Evidence for Differential Localization of Such Sensitivity

Euglena, like many other motile and non-motile eukaryotic unicells and colonial protozoa, has long been known to be sensitive to light and in particular to the blue region of the spectrum. Such a differential sensitivity in *Euglena* has been recorded at least as long ago as 1866 when Cohn[19] noted that *Euglena* accumulated in a region of blue light in apparent preference to red light or dim white light (diffuse sunlight). Engelmann notes[34] that, following illumination of a suspension of *Euglena* with a broad prism spectrum, the cells tended to gather in that region of the spectrum closest to the F (Fraunhofer) line (470–490 nm). Engelmann also notes[34] that passage of *Euglena* from this region into adjacent spectral regions effected a behavioral response similar to that effected on passage of the cells from stronger to weaker light. Oltmans reports[76] similar observations.

Engelmann reports that shadowing of *Euglena* indicates differential sensitivity[34]. A shadow cast upon the anterior end of the cell causes the same behavioral response as casting a shadow over the whole cell and conversely if only the posterior two-thirds of the body are covered with a shadow, no effect is noted (see Fig. 1). Stahl[89] also notes, in passing, that the anterior and posterior ends of *Euglena* show a differential sensitivity to light. Jennings[52] (pp. 137–138) made very careful observations of the behavior of *Euglena* in the vicinity of a light trap. He notes that the cells show a shock reaction on leaving the trap and then notes that "this reaction frequently occurs as soon as the anterior tip is pushed into the shade. In other cases, the reaction does not occur till the *Euglena* has passed completely in the dark; it then turns and passes back into the light". Such reports provide evidence that *Euglena* possesses a blue light photoreceptor and that this photoreceptor is located in the anterior end of the cell. For this review, it would serve little purpose to revive the old arguments concerning the mechanisms by which photobehavior in *Euglena* is controlled, i.e., the stigma photoreceptor versus shading hypothesis[3,25,63,101]. Over the last 70 years, workers have often become engrossed in presenting their own theories or subjective hypothesis to explain the phenomenon, often in deference to actually experimentally analysing the system and providing additional data to prove or disprove their theories. It will suffice then to say that some researchers believe the stigma to be the photoreceptor, while others (and in recent times, the majority) feel that the paraflagellar body is the photoreceptor, with the stigma needed only for accomplishing *oriented* movement through its action as a screening device. We will confine ourselves to presentation of the key results and leave the reader to consult the older literature for a forum on these hypotheses. *Euglena* sp. with both a stigma and the paraflagellar body (PFB)[18,30,44,63,101] as well as bleached (chloropyll free) strains with only a paraflagellar swelling (bleaching induced by drug treatment) and no stigma[15,18,30,36,44] show photobehavior with respect to a light trap. A member of the *Euglena* sp. devoid of both stigma and paraflagellar body, *Astasia*, shows no photobehavior under conditions tested to date[44,59,60,80,81], i.e. does not react to a light trap. It can be concluded from the available reports that the stigma plays a role in determining the threshold levels for photobehavior expresssion, i.e., in the presence of the stigma, cells can show behavior to a reduction in incident light intensity as well as its increase whilst cells without a stigma are only able to respond to increases in light intensity. The reasons for and mechanisms of this sensitization of the cells with a stigma are unknown but, as noted above, have been the subject of much controversy and debate[18,25,31,63,67,101,107,108].

The chlorophyll containing form of *Euglena* which possesses both stigma and paraflagellar swelling, is therefore sensitive to light, in particular that in the blue portion of the spectrum. The anterior end of the cell is most sensitive to light and, in view of correlations that can be drawn between structural features and the expression of photobehavior, we can conclude even from these early observations that the photoreceptor may well be the paraflagellar swelling.

2.2 Action Spectra – Do They Represent the Chromophore Mediating Photobehavior?

As noted above, the earliest study on *Euglena* known to us notes a differential sensitivity of *Euglena* to blue light (Cohn[19]). A careful analysis of the spectral sensitivity of the cell over the entire visible and near-visible portions of the spectrum, might well be expected to provide one with an image of the spectrum of the chromophore mediating the photobehavior. Thus, since the measurements of Engelmann[34], many attempts have been made to record this spectral sensitivity of *Euglena*. The conclusion that can be drawn from 90 years of such studies (a total of 23 action spectra plus 8 other studies on the relative spectral sensitivity at least) is that they have produced rather different results. Most of the studies have one common conclusion – that *Euglena* is particularly sensitive to blue light in a region of 470–520 nm. The spectra are however not all the same (some show peak sensitivity at 410 nm) and are markedly different in fine detail. Both non-polarized and polarized light have been used, the latter providing information on the orientation of pigments involved in photomovement[23]. To however merely discount spectra that do not agree with current opinion would be precipitate for whilst some action spectra have been determined under uncertain conditions with respect to the relative efficiency of incident light (i.e., the energy of incident photons in different portions of the spectrum), even a cursory glance at table 1 shows that uniformity is not exhibited by recent action spectra as well.

The use of an action spectrum analysis to determine the spectral characteristics of a chromophore within a cell that mediates photobehavior has been criticized on the grounds that non-photoreceptor chromophores could severely interfere with such measurements; cf. the analogous problem as discussed in the article by Schmidt (this volume). Indeed, portions of such action spectra of *Euglena* have been discounted by some reviewers on these grounds[73)] (pp. 339). The obvious candidates for such interfering chromophores are the chlorophyll and carotenoid pigments associated with the chloroplasts in green *Euglena* and the cytochromes of the mitochondria. With respect to the former, several workers have turned to the use of chlorophyllfree (bleached) cells[15,18,30,44)] or to dark-grown cells[18,44,109)]. However, the interpretations of such alterations in photobehavior as are effected by either removal of chloroplast pigments or stigma pigmentation in *Euglena* differs. Goessel[44)] and Checcucci et al.[18)] conclude that, due to the similarity of action spectra (of photoaccumulation) in light grown and dark-grown *Euglena*, the chloroplast structures and pigments do not interfere with the determination of the chromophore character of the photoreceptor whilst Diehn and Tollin[30,31)] conclude that, due to a reduction in photoaccumulation activity by dark growth of *Euglena* or administration of inhibitors of photophosphorylation (chloroplast photosynthesis), photophosphorylation is a requirement *per se* for expression of photoaccumulation. All authors[14,15,18,30,36,44)] however agree that the removal of stigma pigmentation by drug-induced (furadantin or streptomycin) bleaching (a process which also effects inhibition of chloroplast pigmentation) effects supression of photoaccumulation to blue light because the presence of the stigma is required for expression of photoaccumulation. The action spectrum of photo-dispersal of such stigma-less cells thus presumably represents the spectral characteristics of the chromophore mediating such behavior in these (altered) cells, especially since, in this case, no other pigment systems are present to obviously interfere with the chromophore of the photoreceptor proper. The presence of other chromophores can however, under some conditions, change the action spectrum of photoaccumulation. As noted by several authors[16,22,109)], *Euglena* will accumulate in a red light trap although even here, the peak spectral sensitivity is not reported to be the same. This red accumulation spectrum is close to that for the chlorophyll absorption and, as generally concluded by these authors[16,108)], the response reflects chemoaccumulation of *Euglena* in an oxygen rich region (trap) – the enhanced O_2 tension being the result of enhanced photosynthetic activity of the cells in the red light trap (see also Sect. 1.7.).

2.3 The Action Spectra: Details and Evaluation

Recent reviews should be consulted for graphical presentation of the action spectra of light trap behavior of *Euglena*. Since sufficient details of either light intensities, spectral output of illumination sources, cell character and previous light history, methods of determination or doses of incident light administered to induce cell behaviors are not given in most published reports of the spectral sensitivity of *Euglena*, it is not possible to either make a direct comparison between the results or to, for example, plot the spectra on common axes for comparison. We will confine ourselves therefore to tabulation of the results obtained over the last 100 years. These results are presented in Table 1 along with a series of footnotes. It should be stressed that such determinations of spectral sensitivity have been made largely without reference to the previous light history of the cells. Some of these measurements were made on cells cultured under light-dark cycles – either natural or artificially imposed[13,14,19,23,24,29,30,34,44,56,57,63,64,76] or on cells cultured under continuous illumination[4,16,17,18,109]. Only in a few of these reports is the light history of the cells, prior to determination of the photobehavior, noted. Buenning and Schneiderhoehn[14] placed their cells in darkness for 24 h before use; Goessel[44] placed cells in darkness for 2 h before use and in more recent studies[4,17,18] cells are placed in darkness for half an hour before use. The effect of such dark adaptation on the threshold sensitivity has not been systematically studied. However, Diehn and Tollin[31] note reduction of photoaccumulation activity of green (light-grown) *Euglena* after prolonged dark incubation (4 h). Wolken and Shin[109] report a marked difference in the spectral sensitivity of light and short-term dark adapted *Euglena.* We therefore have to acknowledge that such treatment could have a significant effect on the relative spectral sensitivity of *Euglena* especially since such sensitivity is determined relative to threshold perception. An elevation in the threshold level could serve to mask weaker responses. A systematic study of the effects of near ultra-violet irradiation (320–370 nm) on *Euglena* motillity has not been reported. The effect of monochromatic ultraviolet light on several protozoa has been reported ([49], pp. 54;[106], pp. 135–138) as well as on *Euglena*[93,96]. The effects range from shock reactions, dispersal of the cells from the illuminated region, and short-term or permanent interuption of ciliary or flagellar activity to (at extreme intensities or doses) bleaching of the cells which arises from impairment of chloroplast synthesis. Since the rate of migration of *Euglena* into a light trap will be in part determined by the overall motility of the cells, photoaccumulation in violet and ultra-violet light traps may be in part caused by a reduction in cell motility once in the trap rather than due to photophobic responses. The reader is reminded of the effects of temperature on the occurence of photoaccumulation ("phototaxis") of *Euglena* ([63], see Sect. 1.7.). The responses of *Euglena* to near-ultra violet light thus remain to be carefully evaluated. Support for this sensitivity appears to come from recent work[4] which reports alterations in swimming behavior following removal of such incident light after very brief illumination. However, the action spectrum in this study is a measure of the relative efficiency of changes in illumination (ΔI) on "activation" of photobehavior within a population of cells rather than for dose/response relationships in individuals cells – only the latter would reveal whether or not the sensitivity is physiological.

Pringsheim[82], in evaluation of the photobehavior spectral sensitivity of *Euglena* proposed that a flavin chromophore mediated photobehavior. Diehn[23,24] following a more exacting study on such spectral sensitivity including dose/response plots for every wavelength postulated that a flavin molecule was responsible for photobehavior although some reviewers ([73], pp. 339) did not agree. Several other recent action spectra[4,17,18] have also been interpreted in the same way, i.e., as supporting the flavin chromophore hypothesis in *Euglena*.

2.4 *In vivo and in vitro Spectroscopic Analysis of Photoreceptive Sites in Euglena*

In pursuit of evidence to either support or disprove the "shading hypothesis", several workers have attempted to determine the in vivo spectroscopic characteristics of the organelles involved in the expression of photobehavior[8,9,10,108] in addition to determining the spectral properties of isolated components or pigments[5,6,8,44,86,88,108]. In vivo absorption spectroscopy measurements of the stigma[8] are different from those taken of that region of the cell in the vicinity of the paraflagellar swelling[107], being reminiscent of the spectra of carotenoids and flavins respectively according to the authors. The high carotenoid content of the stigma granules has been confirmed by both absorption spectroscopy and chromatographic analysis of the isolated granules (globules) and their organic solvent extracts[5,6,86]. In vivo microspectrofluorescence measurements[9,10] reveal that the region of the anterior reservoir of *Euglena* in the vicinity of the paraflagellar swelling emits a yellow-green fluorescence – a fluorescence characteristic that is widely attributed to the presence of flavins for the excitation wavelengths used. However, Sperling-Pagni *et al.*[88] report similar fluorescence characteristics for isolated stigma globules. The evidence thus remains, at this time, as conclusive as the reader chooses to consider but, as noted by Feinleib and Curry[38], characterization of the chromophore in the paraflagellar body requires unequivocable isolation of this organelle from the cell and its in vitro characterization – a preparation that has not yet been reported.

2.5 *Non-Spectroscopic evidence for Flavin vs. Carotenoid Chromophores*

Different photoreceptor pigments may reasonably be expected to undergo different primary photoprocesses upon light absorption, aside from possessing different spectral characteristics. Specifically, light absorption in flavins fairly easily leads to electronic excitation of the molecule to the triplet state, while this process does not readily occur in carotenoids[87].

If flavin triplet excitation were involved in photosensory transduction of an organism, then substances that "quench" this excited state, and thus compete with the normal transduction sequence, should selectively inhibit photomovement in this organism. On the other hand, no effect on such an agent would be expected if the chromophore were a carotenoid.

Diehn and Kint[115] and Mikolajczyk and Diehn[71] demonstrated specific inhibition of photoaccumulation and of the step-down photophobic response of *Euglena gracilis*, respectively, by potassium iodide. Iodide ions are known to de-excite flavins, but do not affect carotenoids[87]. For a description of the analogous experiments with corn coleoptiles, see Sect. 3.2 of the contribution by Schmidt in this volume.

2.6 *Photosensory Transduction*

Since neither dose-response plots for incident light nor the action of any chemical or physical agents on such dose-response relationships and photobehavior in general have been systematically determined, we are left at this time with only subjective speculation as to the nature of the photosensory transduction mechanism(s) in *Euglena*. Several authors[11,25,43,75,97,108] have speculated as to the involvement of various ionic species and gradients in the control/induction of photobehavior in *Euglena*. Such speculation is in loose accord with the nature of any sensory transduction pathway in any eukaryotic sensory system, and may well have provided the reader or research worker with a false impression of our understanding of this process.

3 Summary and Perspectives

Current indirect and direct spectral evidence obtained on *Euglena* indicates that the chromophore mediating photobehavior in this cell possesses characteristics most readily attributed to that of an isoalloxazine ring system. However, it must be emphasized that the determination of these spectral characteristics has been often executed under insufficiently controlled conditions from a spectroscopic, physiological and biochemical standpoint. A quantitative analysis of the transduction sequence should serve to identify the flavin pigment as well as provide details of the primary transductive event which must necessarily, due to the molecular characteristics and reactivity of an isoalloxazine skeleton, be very different from that in a carotenoid sensitized transduction pathway. It can only be hoped that any future studies on photosensory behavior in *Euglena* will be executed with a precision and perspective such that it will not take too long before we have compelling evidence as to the nature of, for example, the primary photocoupling event – elucidating the nature of which would appear to be the next immediate goal in such studies.

Table 1. Photoaccumulation in *Euglena* sp.

Peaks[a]	Spectral range tested (nm)	Spectral source and character[b]	Light source and intensity	Notes cells, type of spectrum/apparatus	Reference
Blue	n.s.	n.s.	microscope light	[0][b]; [LDn][d]; [L-ns][e] Light trap. Visual evaluation	Cohn 19)
Blue[1]	2	Projected prism spectrum	Sunlight or gas light. Dim light	[V]; [LDn]; [L-ns] Prism spectrum. Visual evaluation	Engelmann 34)
458–506[3]	Violet through yellow	Projected prism spectrum. 18–23 cm wide	Carbon arc	[V]; [LDn]; [L-ns] Prism spectrum. Photographic evaluation	Loeb and Wasteneys 57)
475[4]	392–631	Projected prism spectrum through slotted screen	Carbon arc	[V]; [LDn]; [L-ns] Double light beam (trap) method	Loeb and Wasteneys 57)
460–490 402–407[5]	400–600	Projected prism spectrum through slotted screen	Projection lamp of 2000 candles. Dim light[6]	[V]; [LDn?]; [L-ns] Single light trap method. Visual evaluation	Oltmans 76)
483.4[7]	422–524	Constant deviation spectrophotometer	80 W tungsten lamp	[V], [B], [O]; [LDn]; [L-ns]. Double light trap[8]. Red background illumination. Visual evaluation	Mast 64)
425, 450, 476, *495*[9]	400–568	Interference filters with 9–17 nm band width	Projection lamp through microscope 150–750 mW. M^{-2}	[*Gracilis*]; [LDn]; [Ld-24]; Threshold spectra for light intensity to give accumulation of cells 10 min after exposure to light trap. Visual evaluation	Buenning and Schneiderhöhn 14)
410, 450, 475[10]	389–533	Interference filters with 7–17 nm band width	6 V microscope lamp	[M.PBZ-G4 from B][11]; [DD]; [Ld-2]. Light trap. Visual evaluation spectra shows activity of cells (as a function of intensity) after 6 min exposure, relative to response at 410 nm (= 100%).	Gössel 44)

Table 1. Continued

Peaks[a]	Spectral range tested (nm)	Spectral source and character[b]	Light source and intensity	Notes cells, type of spectrum/apparatus	Reference
420, *490*, > 600	388–646	Filter combinations with ~ 40 nm band width	20 W fluorescent tube. 15 ft candles constant incident intensity	[*Gracilis*]; [LL]; [Li]. Double light trap selection method. Relative accumulation after time. Visual evaluation.	Wolken and Shin [109)]
420, 468, 508, > 630	388–646	as above	as above	as above but using plane polarized light	Wolken and Shin [109)]
425, *490*	388–646	as above	as above	[*Gracilis*]; [DD] [Ld]. Other details as above	Wolken and Shin [109)]
490	400–700	Interference filters with 7–17 nm band width	Tungsten lamp	[G]; [LDi]; [L-ns][12] Phototaxigraph (light trap). Rate photo-accumulation. Spectrum is relative response to that at 499 nm corrected for incident quanta	Diehn and Tollin [29)]
375, *475*	350–550	Spectrophotometer grating prism with ~ 10 nm band width	xenon lamp	[G]; [LD:]; [L-ns][12] Phototaxigraph. Rate photoaccumulation. Spectrum is of relative efficiency (as a function of incident intensity) to produce constant response	Diehn [23)]
380, 440, *450*, 480[14]	340–520	Interference filters with ~ 12 nm band width	75 W xenon lamp	[G]; [LL] [Ld-30 min] Phototaxigraph. Rate photoaccumulation. Spectra is plot of ratio between rate photoaccumulation and no. of incident photons.	Checcucci, et al. [18)]
375, *465* 490		As above	As above	As above except [DD]	Checcucci, et al. [18)]

Photodispersal

Peaks[a]	Spectral range tested (nm)	Spectral source and character[b]	Light source and intensity	Notes cells, type of spectrum/apparatus	Reference
440–490	400–600	Projection prism spectrum	Projection lamp of 2000 candles. Bright light[15]	[V]; [LDn]; [L-ns] Light trap. Visual evaluation	Oltmans[76)]
483.4[7]	422–524	Constant deviation spectrophotometer	80 W tungsten lamp	[V]; [LDn]; [Lns][16] Double light trap. Visual evaluation.	Mast[64)]
435, *487* 506[17]	400–540	Interference filters	Lateral illumination of 650–800 Lux	[*Gracilis*]; [LDn]; [Li]; Visual evaluation. Spectrum shows relative sensitivity (as a function of intensity) of cells to migrate away from light compared to response at 410 nm.	Buenning and Gössel[13)]
350, 412 450, 475[13]	350–550	Monochromator ~ 10 nm band band width	150 W xenon lamp	[G]; [LD]; [L-ns][13] Phototaxigraph. Rate Photodispersal using polarized light. Plot inverse of intensities producing constant response.	Diehn[23,24)]
410, 450, 475	389–533	Interference filters with 7–17 nm band width	6 V Microscope lamp	[B][18]; [DD]; [Ld-2] Light trap. Visual evaluation. Spectrum shows relative activity of cells (as a function of intensity) with respect to light source 6 minutes after exposure, relative to response at 410 nm (= 100%)	Gössel[44)]
410, 450	400–455	Interference filters with 9–17 nm band width	as above	[M-PBZ-94 from B][11]; [DD]; [Ld-2]; Other details as above	Gössel[44)]
370, *450*	340–520	Interference filters with ~ 12 nm band width	75 W xenon lamp Up to 8.5 $W.M^{-2}$ incident light	[M-from 9][19]; [LL]; [Ld-30 min]: Phototaxigraph. Rate photoaccumulation. Spectrum is plot of ratio between rate photodispersal and no. of incident photons	Checcucci, et al.[18)]

Table 1. Continued

Photophobic responses

Peaks[a]	Spectral range tested (nm)	Spectral source and character[b]	Light source and intensity	Note cells, type of spectrum/apparatus	Reference
375, 440, 450, 490[14]	360–505	Interference filters with ~ 10 nm band width	75 W xenon lamp. 0–10 mW.M^{-2} incident light	[G]; [LL]; [Ld-30 min]: Evaluation from video film. Spectrum shows relative activating efficiency of light (as a function of intensity) with respect to the percentage of a population of cells that exhibit a tumbling response on removal of light after a constant period of illumination.	Barghigiani, et al.[4)]

a Wavelength of peaks in spectrum. "Peaks of greatest intensity are italicized regardless of whether the spectra are corrected or not."

b Where given, interference filter (or other filter combinations) band widths are for 50% transmittance.

c Symbols denote *Euglena* species (sp.) used:
[G] – *E. gracilis* Z; [B] *E. gracilis* var bacillaris; [gracilis] – *E. gracilis* no other specifications given or var. Pringsheim.; [V] – *E. viridis*; [0] *Euglena* of other species or not specified; [M-] mutant *Euglena*, details given inside brackets and in footnotes.

d Denotes lighting conditions used during growth of the cells: [LDn] – light-dark cycle of natural origin; [LDi] – artifically imposed light-dark cycle in 24 h period; [DD] cells grown in continuous darkness; [LL] – cells grown under continuous illumination.

e Denotes previous light history of the cells, i.e., between removal from culture and use in photobehavior studies:
[L-ns] – light history not specified; [Li] cells adapted to light prior to test; [Ld-2] – cells adapted to darkness for, for example, 2 h, prior to use.

1 69% cells gathered into 470–490 nm region [b5/6F-F^4/7G] and 96.5% within 460–500 nm (wavelength calculations from Mast[64)]).

2 Divided prism spectrum into 6 regions from indigo (violet) to red.

3 Projected a prism spectrum across a trough containing cells. After 30 min cells had accumulated in the region between 410–515 nm; after 3 h cells were found between 458 and 506 nm.

4 Using same prism spectrum, placed slide (containing narrow slits) between prism spectrum and cells. Cell suspension thus illuminated with two light beams (~ 10 nm band width) simultaneously onto a common focussed point. Using one beam constant at ~ 470 nm, investigated the relative effectiveness of other wavelengths to induce cells to migrate out of the blue light trap. No other wavelength was found to be effective.

5 This is the essential character of these spectra although differences (essentially spreading of the peaks) are noted with different light intensities (see Notes 6 and 15). It is of interest to note how Oltmans was prompted to investigate the violet light sensitivity of *Euglena*. He noted that a crude carotene extract from *Euglena* showed absorbance in the blue and violet and, reasoning that the cells should show the same spectral sensitivity in vivo for a carotene photobehavior chromophore, modified his apparatus (new prism) to permit investigation of this.

6 No absolute intensities are given. Notes, however, that with wide slit widths, i.e. high incident light intensity, the cells dispersed from lighted regions and thus he used smoked glass prisms and narrow slit widths. With intermediate slit widths, he notes dispersal of cells from spectral regions in the center of strong photoaccumulation zones, i.e., blue light photodispersal (see Note 15).

7 Spectra corrected for energy distribution of lamp.

8 Two light beams (traps) projected onto cell suspension. One trap was white light (40.6 candles) and the other of monochromatic light. Adjusted the relative intensity of this monochromatic light until the cells swam along a pre-selected path.

9 Reciprocal threshold spectrum, i.e. incident intensity at various wavelengths required to give accumulation. The peaks thus indicate efficacy of that wavelength. Spectra not corrected for energy.

10 Threshold spectra but plotted relative to response of cells to 410 nm light which, being the most effective, was arbitrarily defined as 100%.

11 Bleached (cholophyll free) cells without stigma but possesing paraflagellar body (PFB).

12 Not given in papers but cells generally maintained in light until shortly before experiments upon which the cells were transferred to darkness for 15–30 min (i.e. [Ld-15–30 min]) and then subsequently manipulated under dim red light.

13 Composite relative sensitivity spectrum obtained by combining spectral sensitivity to polarized light aligned parallel to long axis of organisms (violet region of spectrum) and aligned perpendicular to long axis (blue region of spectrum).

14 Peaks at 375 and 450 nm, shoulders at 440 and 490 nm.

15 No spectrum given but notes that with wide slit widths, cells disperse from this region.(see Note 6).

16 Determined for samples that, for undetermined reasons, showed photodispersal response to low intensity light as opposed to photoaccumulation.

17 Shoulder at 505 nm. Cells are green (chlorophyll containing) and possess both stigma and paraflagellar body (as with all normal *Euglena*).

18 Cells grown in dark, i.e., are colorless (bleached – chlorophyll free) but have both stigma and paraflagellar body.

19 Mutant induced with streptomycin treatments; bleached (chlorophyll free) cells without stigma but possessing paraflagellar body.

4 References

1. Aderhold, R.: Zur Kenntnis richtender Kräfte bei der Bewegung niederer Organismen, Jena Z. Naturwissenschaften *15*, 310 (1888)
2. Ascoli, C. et al.: Biophys. J. *24*, 585 (1978)
3. Bancroft, F.W.: J. exptl. Zool. *15*, 383 (1913)
4. Barghigiani, C. et al.: Photochem. Photobiol. *29*, 1015 (1979)
5. Bartlett, C.M. et al.: Plant Physiol. *49*, 881 (1972)
6. Batra, P.P., Tollin, G.: Biochim. Biophys. Acta *79*, 371 (1964)
7. Bendix, S.W.: Botan. Rev. *26*, 145 (1960)
8. Benedetti, P.A. et al.: Arch. Microbiol. *111*, 73 (1976)
9. Benedetti, P.A., Checcucci, A.: Plant Sci. Lett. *4*, 47 (1975)
10. Benedetti, P.A., Lenci, F.: Photochem. Photobiol. *26*, 315 (1977)
11. Bovee, E.C., Jahn, T.L.: J. Theoret. Biol. *35*, 259 (1972)
12. Buder, J.: Jahrbuch Wissenschaftl. Botanik. *58*, 105 (1917)
13. Buenning, E., Goessel, I.: Arch. Mikrobiol. *32*, 319 (1959)
14. Buenning, E., Schneiderhoehn, G.: Arch. Mikrobiol. *24*, 80 (1956)
15. Buenning. E., Tawada, M.: Arch. Mikrobiol. *27*, 306 (1957)
16. Checcucci, A. et al.: Photochem. Photobiol. *19*, 223 (1974)
17. Checcucci, A. et al.: Monitore Zool. Ital. *9*, 83 (1975)
18. Checcucci, A. et al.: Photochem. Photobiol. *23*, 51 (1976)
19. Cohn, F.: Hedwigia *5*, 161 (1866)
20. Colombetti, G., Diehn, B.: J. Protozool. *25*, 211 (1978)
21. Creutz, C., Colombetti, G., Diehn, B.: Photochem. Photobiol. *27*, 611 (1978)
22. Creutz, C., Diehn, B.: Abstr. 2nd Ann. Meet. Amer. Soc. Photobiol., Vancouver, July 1974
23. Diehn, B.: Biochim. Biophys. Acta *177*, 136 (1969)
24. Diehn, B.: Nature (London) *221*, 366 (1969)
25. Diehn, B.: Science (Wash.) *181*, 1009 (1973)
26. Diehn, B.: Photic responses and sensory transduction in motile protists, in *Handbook of Sensory Physiology*, (H. Autrum, ed.), pp. 24–68, Springer Berlin. 1979
27. Diehn, B., Fonseca, J.R., Jahn, T.L.: J. Protozool. *22*, 492 (1975)
28. Diehn, B. et al.: Photochem. Photobiol. *26*, 559 (1977)
29. Diehn, B., Tollin, G.: Photochem. Photobiol. *5*, 523 (1966)
30. Diehn, B., Tollin, G.: ibid. *5*, 839 (1966)
31. Diehn, B., Tollin, G.: Arch. Biochem. Biophys. *121*, 169 (1967)
32. Dryl, S.: Behavior and motor responses of *Paramecium*, in *Paramecium, A Current Survey*, (van Wagtendonk, W.J., ed.), pp. 165–218, Amsterdam: Elsevier 1974
33. Ehrenberg, C.G.: Die Infusionsthierchen als vollkommene Organismen. Leipzig: L. Voss 1838
34. Engelmann, Th.W.: Arch. ges. Physiol. *29*, 387 (1882)
35. Famintzin, A.: Bull. Acad. Imp. Sci. St.-Petersbourg *10*, 534 (1866)
36. Ferrara, R., Banchetti, R.: J. exptl. Zool. *198*, 393 (1976)
37. Feinleib, M.E.: Ph.D. thesis, Harvard Univ. Cambridge Mass. (1965)
38. Feinleib, M.E., Curry, G.M.: The nature of the photoreceptor in phototaxis, in *Handbook of Sensory Physiology. Vol. I. Principles of receptor physiology,* (W.R. Loewenstein, ed.), pp. 366–395. Berlin: Springer 1971
39. Forward, R.B. jr.: Planta *111*, 167 (1973)
40. Forward, R.B. jr.: Protozool. *21*, 312 (1974)
41. Fox, H.M.: J. Gen. Physiol. *3*, 483, 501 (1921)
42. Frank, T.: Botan. Zeitung *62*, 153 (1904)
43. Froehlich, O., Diehn, B.: Nature (London) *248*, 802 (1974)
44. Goessel, I.: Arch. Mikrobiol. *27*, 288 (1957)
45. Gunther, F.: Archiv. Protistenk. *60*, 511 (1927)
46. Halldal, P.: Physiol. Plant. *11*, 118 (1958)

47. Halldal, P.: Phototaxis in Protozoa, in *Biochemistry and Physiology of Protozoa.* Vol. III., (Hutner, S.H. ed.), pp. 277–296. New York: Academic Press 1964
48. Jahn, T.L., Bovee, E.C.: Locomotive and motile response in *Euglena,* in *Biology* of *Euglena.* (Buetow, D.E., ed.), pp. 45–108. New York: Academic Press 1968
49. Jahn, T.L., Bovee, E.C.: Locomotive and motile behavior in Protozoa, in *Research in Protozoology*. Vol. I. (Chen, T.T., ed.), pp. 47–200. Oxford: Pergamon Press 1968
50. Jennings, H.S.: Am. J. Physiol. *2*, 355 (1899)
51. Jennings, H.S.: *Contributions to the study of the behavior of lower organisms*, Carnegie Inst. Washington, D.C. (1904)
52. Jennings, H.S.: *Behavior of Lower Organisms*, Columbia Univ. Press. New York (1906)
53. Jennings, H.S., Moore, E.M.: Am. J. Physiol. *6*, 233 (1903)
54. Leedale, C.F.: *The Euglenoid Flagellates*, Englewood Cliffs, N.J.: Prentice Hall 1967
55. Lindes, D., Diehn, B., Tollin, G.: Rev. Sci. Instr. *36*, 1721 (1965)
56. Loeb, J., Maxwell, S.S.: Univ. Cal. Publ. Physiol. *3*, 195 (1910)
57. Loeb, J., Wasteneys, H.: J. Exptl. Zool. *20*, 217 (1916)
58. McNab, R., Koshland, D.E. jr.: J. Mol. Biol. *84*, 399 (1974)
59. Mainx, F.: Arch. Protistenk. *60*, 305, 355 (1927)
60. Mainx, F.: Arch. Protistenk. *68*, 105 (1929)
61. Massart, J.: Arch. Biologie. *9*, 515 (1889)
62. Mast, S.O.: J. Exptl. Zool. *3*, 359 (1906)
63. Mast, S.O.: *Light and the behavior of Organisms*, New York: Wiley 1911
64. Mast, S.O.: J. Exptl. Zool. *22*, 471 (1917)
65. Mast, S.O.: J. Exptl. Zool. *26*, 503 (1918)
66. Mast, S.O.: Motor responses to light in the invertebrate animals, in *Biological Effects of Radiation.* (Dugger, B.M., ed.), pp. 573–623 (1936)
67. Mast, S.O.: Motor responses in unicellular animals, in *Protozoa in Biological Research.* (Calkins, G.N. and Summers, F.M., eds.), pp. 271–351. New York: Columbia Univ. Press 1941
68. Mast, S.O., Glover, M.: Biol. Bull. *43*, 203 (1922)
69. Mast, S.O., Johnson, P.L.: Z. Vergl. Physiol. *16*, 252 (1932)
70. Merton, H.: Biol. Zentralbt. *55*, 268 (1935)
71. Mikolajczyk, E., Diehn, B.: Photochem. Photobiol. *22*, 269 (1975)
72. Mohri, H.: Biochem. Biophys. Acta *456*, 85 (1976)
73. Naitoh, Y., Eckert, R.: Control of Ciliary Activity in Protozoa, in *Cilia* and *Flagella* (Sleigh, M.A., ed.), pp. 305–352. London: Academic Press 1974
74. Nultsch, W., Throm, G., Rimscha, I.V.: Arch. Microbiol. *86*, 351 (1971)
75. Nultsch, W., Haeder, D.P.: Photochem. Photobiol. *29*, 423 (1979)
76. Oltmans, F.: Z. Botanik *9*, 257 (1917)
76a. Pietrowicz-Kosmynka, D.: Chemotaxis of *Stentor corueleus* in dependence on the ionic composition of medium. Abstr. 3rd Int. Congr. Protozool. Leningrad, pp. 155 (1969)
77. Poff, K.L., Loomis, W.F. jr.: Exptl. Cell Res. *82*, 236 (1973)
78. Poff, K.L., Loomis, W.F. jr., Butler, W.L.: J. Biol. Chem. *249*, 2164 (1974)
79. Pringsheim, E.G.: Planta, *2*, 555 (1926)
80. Pringsheim, E.G.: Cytologia, Fujii Jubilaeumsband, pp. 234–255 (1937)
81. Pringsheim, E.G.: New Phytol. *47*, 52 (1948)
82. Pringsheim, E.G.: Nova Acta Leopoldina. Abhandl. Dtsch. Akad. Naturforscher (Leopoldina). *125*, (18) 1 (1956)
83. Robertson, J.A.: Arch. Microbiol. *85*, 259 (1972)
84. Schletz, K.: Z. Pflanzenphysiol. *77*, 189 (1976)
85. Sleigh, M.A. (ed.) *Cilia* and *Flagella.* London: Academic Press 1974
86. Sperling-Pagni, P.G. et al.: J. Phycol. Suppl. *9*, 20 (1973)
87. Song, P.S., Moore, T.A.: J. Am. Chem. Soc. *90*, 6507 (1968)
88. Sperling-Pagni, P.G., Walne, P.L., Wehry, E.L.: Photochem. Photobiol. *24*, 373 (1976)
89. Stahl, E.: Botanische Z. *38*, 409 (1880)
90. Stavis, R.L., Hirschberg, R.: J. Cell Biol. *59*, 367 (1973)

91. Strassbourger, E.: Jena Z. Naturwiss. *12*, 551 (1878)
92. Summers, K.E.: Biochim. Biophys. Acta *416*, 153 (1976)
93. Swann, W.F.G., Del Rosario, C.: J. Franklin Inst. *213*, 549 (1932)
94. Tartar, V.: Exptl. Cell Res. *13*, 317 (1957)
95. Taylor, B.L., Koshland, D.E. jr.: J. Bacteriol. *123*, 557 (1975)
96. Tchakhotine, S.: Compt. Rend. Soc. Biol. Paris. *121*, 1162 (1936)
97. Tollin, G.: *Curr. Topics Bioenergetics, 3*, 417, New York: Academic Press 1973
98. Tollin, G., Robinson, M.I.: Photochem. Photobiol. *9*, 411 (1969)
99. Torrey, H.B.: Science (Wash.) *37*, 873 (1913)
100. Treviranus, L.C.: Fernere Beobachtungen über die Bewegung der grünen Materie im Pflanzenreiche, in *Vermischte Schriften anatomischen und physiologischen Inhalts von G.C. und L.C. Treviranus, 2*, 79 (1817)
101. Wager, H.: J. Linnean Soc. (London) Zool. *27*, 463 (1900)
102. Wager, H.: Phil. Trans. Roy. Soc. London Ser. B. *201*, 333 (1911)
103. Walker, E.B., Lee, T.Y., Song, P.S.: The *Stentor* photoreceptor. Abstr. 7th Ann. Meet. Amer. Soc. Photobiol. Asilomar/CA (1979)
104. Watanabe, M., Miyoshi, Y., Furuya, M.: Plant cell Physiol. *17*, 683 (1976)
105. Watanabe, M., Furuyama, M.: Plant Cell Physiol. *15*, 413 (1974)
106. Wichterman, R.: *The Biology of Paramecium.* Blakiston, New York (1953)
107. Wolken, J.J.: *Euglena. An experimental organism for biochemical and biophysical studies.* Rutgers Univ. Press (1967)
108. Wolken, J.J.: J. Protozool. *24*, 518 (1977)
109. Wolken, J.J., Shin, E.: J. Protozool. *5*, 39 (1958)
110. Wood, D.C.: Photochem. Photobiol. *24*, 261 (1976)
111. Worley, L.G.: J. Cell Comp. Physiol. *5*, 53 (1934)
112. Holwill, M.E.J.: J. Exptl. Biol. *44*, 759 (1966)
113. Piccinni, E., Albergoni, V.: J. Protozool. *20*, 456 (1973)
114. Nichols, K.M., Rikmenspoel, R.: J. Cell. Sci. *29*, 233 (1978)
115. Diehn, B., Kint, B.: Physiol. Chem. Phys. *2*, 483 (1970)

Sensory Physiology of Phycomyces Blakesleeanus

Vincenzo E. A. Russo and Paul Galland

Max-Planck-Institut für Molekulare Genetik, Abt. Trautner, Ihnestr. 63–73, D-1000 Berlin 33, Federal Republic of Germany.

Table of Contents

The article describes the present status of the research with the fungus *Phycomyces blakesleeanus* and presents an analysis of the sensory reaction such as phototropism, light-growth response autochemotropism and photodifferentiation. Emphasis is laid on the usefulness of a genetic approach to unravel the light transduction pathway of short term and long term light responses. The advantages of *Phycomyces* as a model system for blue light reactions and especially its potentials in studying range adjustment are outlined.

A. Introduction

Phycomyces has aroused the interest of biologists for over a century. Yet, in the present mode of approach and definition of its aims, the research with this fungus, as it is being conducted nowadays, is clearly a child of the early period of molecular biology.

In biology, perhaps more than in any other science, the success in solving fundamental problems is intimately linked to the choice of a suitable object displaying the phenomenon under study. This has never been more obviously demonstrated than by the eminent success of the early work with bacteria and bacteriophages resulting in the deciphering of the genetic code and proving its universality. It is clear that whereever possible other complex properties of life, such as differentiation and signal processing, should be studied in simple organisms which are amenable to genetical analysis. Simplicity, as such, is, however, of no value if the functions of the analyzed organism do not represent universally represented mechanisms. The existence of universal valid mechanisms may itself be questioned, when one considers the tremendous multitude of signal perception and signal processing. In fact, when Delbrück and Reichardt (1956) started their research on Phycomyces, it was not the intent to introduce a universally valid model system. *Phycomyces* was chosen as a suitable experimental tool because the graded, i.e., linear responses of the fungus seemed easier to analyse than all or none, i.e., nonlinear responses as displayed in nerve excitation. Our knowledge of the differentiation and perception-reaction mechanisms is still at such a rudimentary stage that a quest for a universal mechanism of signal processing – if such a thing exists – would appear premature, at least before the phenomenon is analyzed in detail for some simple cases.

Unfortunately, the criteria for the simplicity of a model organism are unclear. A low evolutionary position and lack of morphological sophistication can be of great technical advantage in the laboratory, but they may be poor criteria with respect to the real simplicity of the organism. Indeed, considering the high complexity of sensorial perceptions to which *Phycomyces* responds serious doubts may arise in the mind of some readers as to the simplicity of this system. Must not the unicellular sporangiophore of *Phycomyces*, capable of reacting to light, gravity, gases and stretch, possess much more complicated mechanisms than the multicellular organs of evolutionary higher organisms, which respond to just one signal? Is not the morphological and technical simplicity of the sporangiophore outweighed by the complexity of its reactions?

We will show in this review that this is not the case. The many different input signals are using the same output system; furthermore it can be shown, that in combination with an analysis of behavioral mutants, that the reactivity to many different stimuli turns out to be of great help in the genetic dissection of the signal-effector pathway. The isolation and genetical characterization of behavioral mutants, which are defective in phototropism has been extremely valuable in this analysis. Physiological characterization of these mutants is still in progress and attempts are being made to isolate new mutants.

An important basis for the choice of *Phycomyces* as a model system is its wide range of sensitivity to light which covers about nine decades and which is therefore similar to vertebrate vision. This makes the organism especially suitable for the study of range adjustment, i.e., the phenomenon of adaptation. The potential of *Phycomyces* as a model system for investigation of differentiation and photo-differentiation has not yet been fully exploited.

Studies of morphogenesis and photoinduction of β-carotene began late in this decade as the desire to unravel the mechanism of the light-growth response and phototropism dominated for many years. We would like to demonstrate here that *Phycomyces* offers itself well as an object for the study of photomorphogenesis. Today, to the best of our knowledge *Phycomyces* seems to be, the only organism in plant physiology which permits a ,,vertical" analysis of both short- and long-term light reactions. The mutants available should be of great advantage since they permit an analysis of the extent to which both kinds of reactions use common steps in light signal processing.

In this review we attempt to give a general introduction to the sensorial physiology of *Phycomyces* omitting the majority of the genetic studies which are the subject of a recent review (Cerdá-Olmedo, 1977). Detailed aspects of the biology of the fungus and an extensive discussion of older literature have been reviewed by Bergman et al., (1969).

B. The Biology of *Phycomyces Blakesleeanus*

Phycomyces belongs to the order *Mucorales*. Its asexual life cycle is characterized by the germination of spores which enlarge, when put on a suitable medium, and form one or two germ tubes, which in turn proliferate into a radially growing mycelium. After 50 to 70 h the mycelium forms giant aerial hyphae, the sporangiophores, which in the mature stage develop a sporangium, containing 10^4 to 10^5 spores. *Phycomyces* is a coenocynot, i.e., it is a single celled organism containing many thousands of nuclei; there is no compartmentalization inside the mycelium or inside the sporangiophore between nuclei. The spores are ellipsoid cells with a thick wall and contain 2 to 4 nuclei (Reau, 1972). Good germination induction of the spores requires a heat shock at 50 °C.

The sporangiophores are straight, fast growing (about 1 mm/h) aerial hyphae of 0.1 mm diameter. The tip is yellow due to accumulation of β-carotene (Stage I). When the sporangiophores reach a length of 1 to 2 cm the tip of the sporangiophore enlarges considerably (Stage II) and forms a sporangium of 0.5 to 0.8 mm diameter. At this stage the newly formed sporangium is a brilliant yellow (β-carotene). At later stages the sporangium color changes to almost black. After sporangium formation the sporangiophores stop growing for several hours (Stage III) before the fast growth is again resumed. Subsequently a 90 min period of elongation occurs which is accompanied by counterclockwise rotation of the sporangium (Stage IVa); thereafter the direction of rotation changes to clockwise (Stage IVb). Clockwise rotation remains during the remainder of the lifetime of a sporangiophore which can grow to a length of 10–20 cm (Fig. 1). Physiological experiments are almost exclusively performed with Stage IVb sporangiophores of 2.5–3.5 cm length. In this stage the average growth velocity is 3 mm/h.

Elongation of the sporangiophore is confined to a small sensitive zone, the so-called growing zone, beginning 0.1 to 0.5 mm below the sporangium and extending downwards 2 to 3 mm. This growing zone can again be subdivided with respect to maximal stretch and maximal twist. In the upper part stretch is predominant and decreases downwards to a region where twist becomes maximal (Cohen and Delbrück, 1958). The growing zone is transparent and cylindrical with a large vacuole in its center. The vacuole extends along the whole length of the sporangiophore. The cytoplasm between cell wall and vacuole is 10 to 20 μm thick. During growth the growing zone does not change its dimensions and this means that the region of the growing zone distal to the sporangium is continuously being converted into insensitive wall material and that the material immediately below the sporangium grows into the growing zone and becomes sensitive material which can subsequently stretch and twist. The pressure inside the sporangiophore is 2 atm (30 psi) (Roelofsen, 1950) which keeps the fragile growing zone straight.

The growth rate of individual organisms is very constant and varies less than 10% after a period of many hours. The growth rate of different organisms from the same population can, however, vary from 40 to 80 μm/min. The sporangium rotates with a speed of about 12°/min.

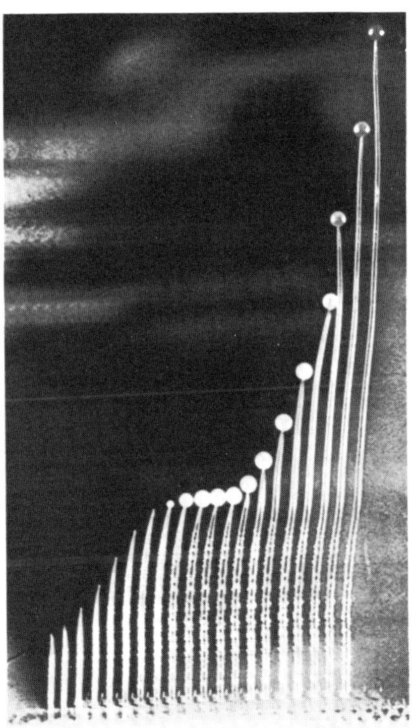

Fig. 1. Growth stages of the sporangiophore. The same sporangiophore was photographed at hourly intervals. (Photo Lois Edgar)

C. Description of the Various Responses

1. Light- and Dark-Growth Responses

Sporangiophores respond to changes of light intensity with a transient change of the growth rate. A short pulse of light of higher intensity than the environmental light causes a transient increase in the growth rate which reaches a maximum four minutes after the pulse. The growth rate then drops for some minutes below the normal rate before the normal rate is reestablished. A corresponding pulse of decreased light causes a transient reduction of the growth rate.

2. Phototropism

Unilaterally illuminated sporangiophores bend towards the light source. If the sporangiophore is in equilibrium with the outside light intensity then the reaction starts with a delay of about 5 min and bending proceeds with a steady rate of ca. 3°/min until an equilibrium between the light stimulus and the upward growth stimulus of gravity is reached.

Phototropism depends on differential light intensities at the proximal and distal side of the sporangiophore relative to the light source. The transparent growing zone has lens properties by which the light received by the proximal side (near the light source) is focused on the distal side. This produces a small band of light of about 10-fold higher intensity on the distal side of the sporangium as compared to the proximal side (Fig. 2). If the sporangiophore is submerged in paraffinoil of the same refractive

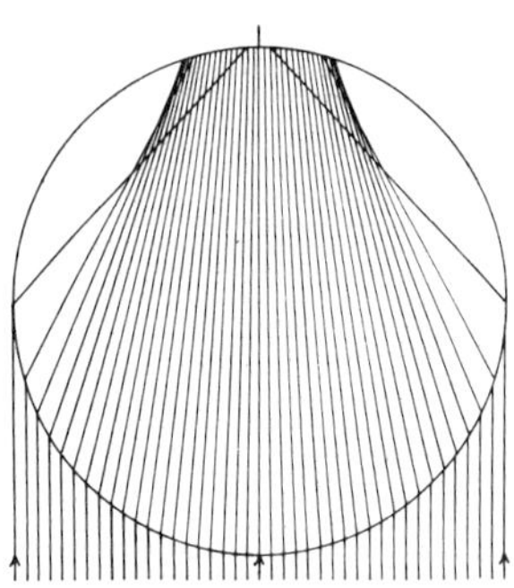

Fig. 2. Diagram of the calculated light paths within the cell's cross section when irradiated in air by a unilateral beam of normally incident parallel light. (From Castle, 1965)

index as the cytoplasm focusing is abolished and no phototropism occurs demonstrating the importance of the lens effect in phototropism (Buder, 1918; Shropshire, 1962). Ultraviolet light (wavelength below 300 μm) causes negative phototropism (Curry and Gruen, 1957) because sporangiophores contain gallic acid which absorbs UV light so strongly that the focus effect is overcompensated (Dennison, 1959; Delbrück and Shropshire, 1960).

3. Geotropism

If sporangiophores are placed horizontally they start to grow vertically by reacting to the force of gravity. This reaction is slower than phototropism and has a delay of 30–180 min.

4. Autochemotropism

Autochemotropism is also known as the avoidance response. Sporangiophores which are placed near an obstacle, e.g., a glass or metal wall, a wire or a silk thread, bend away from these objects. A closely related response to autochemotropism is the so-called house response. If a sporangiophore is covered with a "house", e.g., a cuvette, the specimen responds with a transient increase of the growth rate similar in quantity and duration to the light growth response (Cohen et al., 1975).

5. Anemotropism

Anemotropism is also known as wind response. Sporangiophores which are exposed to a unilateral movement of air bend into the wind (Cohen et al., 1975).

6. Stretch

Mechanical bending of the sporangiophore in the growing zone can be achieved by centrifugation. Stretch occurs on one side of the sporangiophore and relaxation on the other side. Sporangiophores react to relaxation with a transient enhancement of the growth rate and to stretch with a diminution of the growth rate (Dennison, 1961).

7. Growth Response to Various Gases

Ethylene and ethane cause transient growth responses (Russo et al., 1977). Addition of carbon dioxide or diminution of the oxygen concentration (from 20% to 10%) trigger a transient diminution of the growth rate (unpublished results).

8. Photodifferentiation

Various differentiation processes in the vegetative life cycle of *Phycomyces* can be controlled by blue light. Light stimulates the biosynthesis of β-carotene (photocarotenogenesis) and under certain conditions it triggers the initiation of sporangiophores (photophorogenesis) and the formation of sporangia (photosporangiogenesis).

D. Genetical Dissection of the Sensory Transduction Chain

The sensory capacities of the sporangiophore listed above clearly demonstrate the extent to which a seemingly simple cell can perform complex reactions. A genetical approach, to help unravel the complex network of the different *Phycomyces* reactions determined by physiological experiments has been initiated. The phototropic defective mutants isolated fall into two phenotypic classes: Class I mutants have a phototropic threshold raised by a factor of 10^3 to 10^5 ("nightblind mutants") whereas Class II mutants have a normal threshold but a reduced efficiency of bending ("stiff mutants") (Bergman et al., 1973). The phototropic threshold of some of these mutants is depicted in Fig. 3. Complementation analyses assigned these mutations to seven different genes, called *madA* to *madG* (Ootaki et al., 1974; Eslava et al., 1976). The availability of phototropic defective mutants raised the question of which of the other sensory reactions as described above are still functional or defective in the mutants. Assuming that the process of phototropism as a linearly arranged chain of biochemical reactions leading from the light excitation of the receptor pigment to the growth of the cell wall then one can interpret the mutants as having blocks along this chain. It is possible to order them according to their phenotypic capacity to react to various stimuli. Mutants *madA*, *madB*, and *madC* are normal in geotropism and autochemotropism, whereas mutants *madD* to *madG* are defective; one has therefore to place mutants *madD* to *madG* behind mutants *madA*, *madB*, and *madC*. Taking into account further reactions such as photocarotenogenesis, photophorogenesis and photosporangiogenesis one reaches a more detailed model as depicted in Fig. 4. Mutants *madA* and *madB* can be placed before *madC* because of their defect in photophoro-

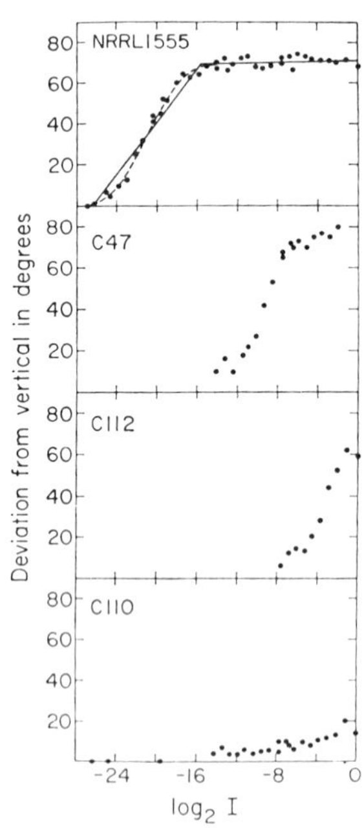

Fig. 3. Phototropism in the wild type NRRL 1555 strain and three mutant strains of *Phycomyces*. After 6 h of illumination with blue light the deviation from the vertical was determined. C47 and C112 are class I mutants; C110 is a class II mutant. (From Bergman et al., 1973)

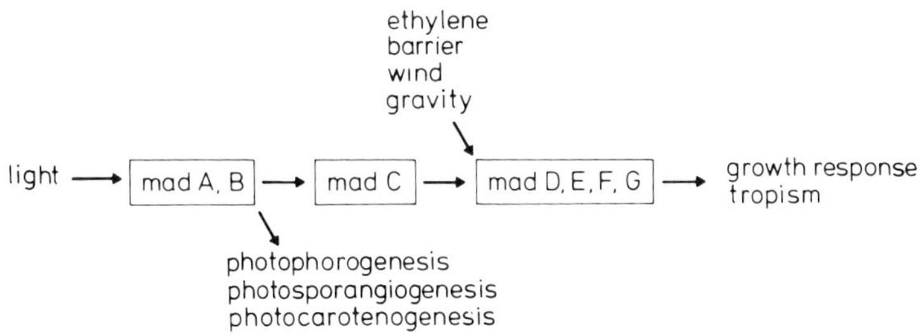

Fig. 4. Pleiotropism of *Phycomyces* strains selected as phototropic defective mutants. Mutants *madA* and *madB* are defective in *all* photoresponses, mutants *madD*, *E*, *F*, *G* only in tropic and growth responses

genesis and photosporangiogenesis (Bergman, 1972; see Sect. F1 and F2). A defect of *madB* mutants in photocarotenogenesis has been reported (Bergman et al., 1973) but no published data are yet available about photocarotenogenesis in *madA* and *madC* mutants.

E. Analysis of Short Term Responses

1. Geotropism

Geotropism is the least studied and the correspondingly least understood reponse of all. The probable reason for this is the difficulty encountered when detailed kinetic studies are undertaken. This is due to the very poor reproducibility of the delay of the response (30 to 180 min) and to the slowness of bending, i.e., 0.07° to 0.3°/min. There have, however, been some interesting results reported. The first is the observation of Pilet (1956) that Stage I sporangiophores have a geotropic bending speed 6 times greater than Stage IV sporangiophores although the growth rate in this stage is 3 times slower than the growth rate of Stage IV sporangiophores. This is in contrast to the phototropic reaction, in which the binding rate increases with increasing growth rate (Shropshire, 1971). It is possible that the structure of the two stages is much more important for geotropism than for phototropism. For example, the "gravity sensor" could sediment better in a Stage I tip (the sensitive part of the sporangiophore) because of the lack of the huge vacuole found in the Stage IV growing zone.

The "gravity sensor" is unknown in *Phycomyces*. Statoliths have not been detected. Any particle with a density different from that of the average density of the cytoplasm could be a candidate, e.g., mitochondria, lipid droplets, nuclei or crystals. The work of Dennison (1961) has clearly shown that the "gravity sensor" must be inside the cell because the response is independent of the density of the medium outside the sporangiophore.

Even before the physiology of *Phycomyces* was approached genetically it was clear that the two responses, phototropism and geotropism, could have a common biochemical pathway (Pilet, 1956; Dennison, 1964). These authors show that blue light inhibits the geotropic response at an intensity much smaller than the intensity which is needed for saturation of the phototropic response. There is, however, no biochemical explanation for the light inhibition of geotropism. What renders this phenomenon more puzzling is the fact that light at the same intensity does not inhibit autochemotropic bending (Cohen et al., 1975; unpublished results) although autochemotropism requires the same gene products, *madD* to *madG*, as required for geotropism. The pleiotropic nature of mutants in genes *madD* to *madG* has confirmed that geotropism and autochemotropism must have a common biochemical pathway (see Sect. D).

2. *Autochemotropism*

If a barrier is placed near the growing zone of Stage IV sporangiophores it grows away from the barrier after a delay of 3–4 min. The bending rate is an inverse function of the distance of the barrier from the growing zone (Lafay et al., 1975). At a distance of 1 mm the bending rate is 3–4 degrees/min.

Other related phenomena are: a double barrier will induce a transient growth response; in still air (very small house 2.5 · 2.5 · 7.5 cm) there is no avoidance response; the avoidance response is restored if an air current is artificially created in this small house (Johnson and Gamow, 1971).

Although the avoidance reaction of *Phycomyces* has been known for almost a century (Elfving, 1881), only recently were Johnson and Gamow (1971) able to present a formal model to explain how the sporangiophore can sense an obstacle and grow away from it. Their model suggests that the sporangiophore emits an unknown gas which is a growth promotor. The obstacle inhibits the random movement of air on the side of the sporangiophore next to the obstacle. As a result more gas accumulates on this side of the sporangium than on the opposite side. The gradient in the concentration of the gas so produced will create differential growth due to the growth inducing capacity of the gas. This differential growth causes bending.

Cohen et al., (1975) have supported their model by demonstrating that there is no avoidance response if a stream of air is blown longitudinal to the sporangiophore. In addition these authors reported another response which they called rheotropism based on the observation that a sporangiophore in a wind tunnel with wind speed of 10 to 150 cm/s bends into the air current at a speed of 1 to 3 degrees/min. This effect can be explained by the Johnson-Gamow model by assuming that the air stream concentrates the growth promoting gas on the downwind side of the sporangiophore. Cohen et al., (1975) have excluded electrostatic fields, humidity, mechanical stretch, and temperature as possible signals mediating the avoidance response. A positive approach to identifying the chemical nature of this unknown gas came from the work of Russo et al., (1977), and Russo (1977a). They suggested a series of criteria for the identification of the gas. These criteria based on the physiological experiments reported above and on the genetical data of Sect. D are the following: (a) addition of gas X above the naturally produced level should give a transient growth response, (b) the avoidance response will be inhibited in the presence of excess gas X, (c) gas X must be produced by Stage IV sporangiophores, (d) mutants in the genes *madD, E, F* (*madG* was not known at that time) should have no gas X induced growth response, (e) these same mutants should make the same amount of gas X as wild type *Phycomyces*. The gases tested were ethylene, acetylene, ethane, carbon dioxide, and nitrogen. Ethylene and ethane satisfied the first two criteria, however, only ethylene satisfied all criteria.

It was therefore concluded that ethylene is the gas X postulated by Johnson and Gamow. The concentration necessary to cause 50% inhibition of the bending rate is 30 ppm, which although low is not as low as the concentration of ethylene which effects higher plants [0.1 ppm in pea growth inhibition, Abeles (1973)].

Phycomyces is by no means the only organism that can regulate its own growth with volatile substances. Bonner and Dodd (1962) have reported that *Polysphondilium pallidum, Dictyostelium discoideum, D. mucoroides, D. purpureum* have the capacity to avoid obstacles. A series of experiments has clearly shown that this orientating effect is due to a gas made by the organism. Attempts to identify this gas have been unsuccessful.

3. Light- and Dark-Growth Responses

a) Concept of Level of Adaptation

The sporangiophore of *Phycomyces* in Stage IV grows at a speed of about 50 μm/min. This growth rate is completely independent of light intensities ranging from dark to 1 W/m². However, a change of light intensity in this range will induce a transient change of growth rate. The light program can be either a pulse of light or a step change. In Fig. 5 the four basic illumination programs and their outputs are shown. The impor-

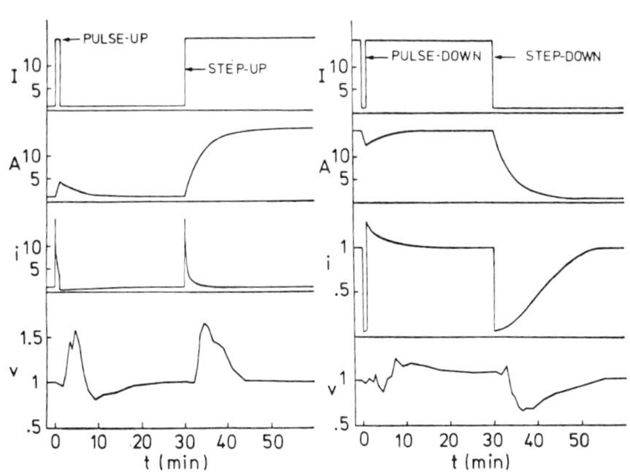

Fig. 5. Four basic illumination programs and their outputs. The top row gives the program, intensity I (linear scale) versus time t. The bottom row gives the growth output, velocity (relative to average velocity) versus time. The second and third row give the level of adaptation A, and the subjective intensity, i = I/A, calculated according to the theory developed. Note that the scale used to plot i(t) is twenty times larger for the "down" than for the "up" programs. (From Delbrück and Reichardt, 1956)

tant features are: (i) with a pulse of light there is first an increase and then a decrease of growth speed before reaching the base-line level, (ii) with a step-up (or step-down) there is only a transient increase (or decrease) in growth rate. In both cases the basal growth rate is reached after less than 30 min. Our discussion will concentrate on the most thoroughly studied situation, namely the pulse-up experiment. The response, R, to a stimulus is defined as the peak to valley amplitude of the growth rate after the pulse stimulus.

Stimulus, S, is defined as

$$S = (I - I_0)\,\Delta t/I_0 \tag{1}$$

where I_0 is the light intensity at which the sporangiophore was pretreated for a long time (h), I is the intensity of the pulse, and Δt is the duration of the pulse (generally less than 1 min). R is then a function of S and not of I, I_0, or Δt separately. This is true for Δt in the range of 1 to 64 s and I_0 in the range of 10^{-4} to 10^{-1} W/m^2 (Delbrück and Reichardt, 1956). The function is

$$R = R_0\, S/(S + S_0) \tag{2}$$

with R_0 and S_0 being constants for a given wavelength, the response has its maximum for $S \geqslant 10^2$ min (Foster and Lipson, 1973). For $S = 10^2$ min the response R is almost independent of the absolute value of I_0 in the range from 1 to 10^{-7} W/m^2 (Fig. 6). The fact that R is a function of S and not of I and Δt alone implies that the same

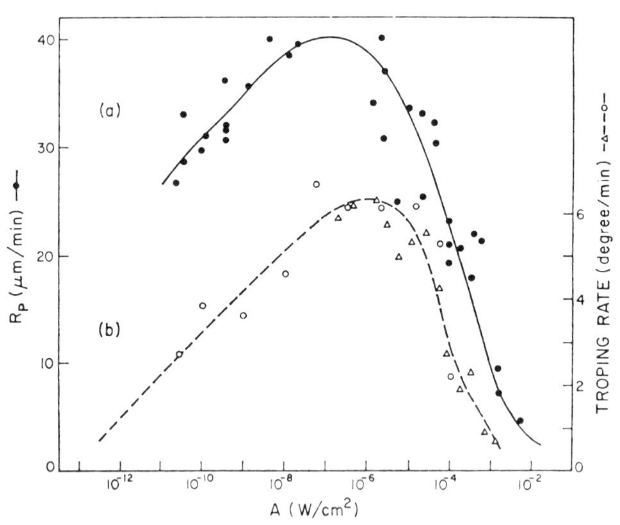

Fig. 6a. Response amplitude Rp̂ (peak-to-peak) as a function of the pretreatment light intensity A for fixed stimulus size $S = 10^2$. **b** Phototropic bending rate as a function of light intensity A. The open triangles are from Bergman et al. (1969). The open circles from Reichardt and Varjú (1958). (From Foster and Lipson, 1973)

pulse of light, with the same number of photons, can either cause the maximal response or no response at all depending on the two conditions:

$I_0 \ll I$ or $I_0 = I$.

Another way to express this fact is to say that the sensitivity of the system changes with the light intensity of the pretreatment. This led Delbrück and Reichardt (1956) to introduce the concept of level of adaptation, A. The level of adaptation, A, is postulated to be a biophysical or biochemical status of the cell and its magnitude is somehow proportional to the intensity of light used in the pretreatment. They assumed that after a long pretreatment at constant intensity the adaptation reaches an equilibrium with intensity, formally $A = I_0$.

Furthermore, after a change of light intensity, A will be described by the differential equation

$$\frac{dA}{dt} = \frac{I - A}{b}$$

where b is the time constant of the system. Finally, the growth is a function of the subjective light intensity $i = I/A$ at any given time.

b) Predictions and Limits of the Delbrück–Reichardt Model

Delbrück and Reichardt tried with a simple and elegant mathematical formulation to explain two very different effects which are caused by changes of light intensity:
(1) the change of level of adaptation (adaptation output),
(2) the change of growth rate (growth output).
Their model predicts the kinetics of the level of adaptation, A, to any change of light. Figure 5 shows the quantitative predictions for the four basic light programs. Another prediction is that the time constant of adaptation in a light step-up or step-down must be identical. And finally the kinetics of the change of the light sensitivity and for the return of the growth rate to the baseline level should be identical.

If we substitute S from formula (1) into formula (2) and put A in place of I_0 we obtain

$$R = \frac{R_0 (I - A) \Delta t}{(I - A) \Delta t + A S_0} . \qquad (3)$$

With the help of formula (3) we can calculate the value of A at any time by measuring the growth response R to a pulse of light of intensity I and duration Δt. For experimental reasons one chooses light pulses that give half-maximal response. To measure the kinetics of dark adaptation the sporangiophore is bilaterally illuminated for a long time at a light intensity I_0. At time 0 the light is switched off. At different times suitable pulses of light are given and the response R is measured. A is then calculated with the help of formula (3). To measure the kinetics of light adaptation the sporangiophore is bilaterally illuminated for a long time at an intensity I_0. At time 0 the light intensity is raised from I_0 to I, at time t the light is switched off and now the kinetics of dark adaptation are measured. By extrapolation of the kinetics of dark adaptation to time, t, it is possible to calculate the level of adaptation A at time t. Following a series of experiments where t is varied it is possible to determine the kinetics of light adaptation, A, after a step-up of light. The kinetic of dark adaptation was indeed shown to be exponential as predicted and the time constant *b* calculated to be either 4 min (Delbrück and Reichardt, 1956) or *b* = 6 min (Lipson and Block, personal communication).

The time constant *b* in light adaptation after a small step-up was found to be 5 min by Delbrück and Reichardt (1956). Lipson and Block (personal communication) found a much faster kinetic following a larger step-up. This problem must be further investigated to determine if the kinetic of light adaptation obeys the Delbrück and Reichardt model only for small stimuli. The decay of A is exponential for at least 50 min (Lipson and Block, personal communication) and in the phototropic response is exponential for at least 90 min (see Sect. E. 4. b).

The growth rate, however, always returns to the baseline level less than 30 min after the stimulus as shown in Fig. 7. This observation demonstrates that the model is not correct for the growth response as there must be at least two biochemical steps along the light sensory transduction chain which are subject to regulation. One sets the level of sensitivity and the other is responsible for the return of growth rate to the base-line level. Only the first step which sets the level of sensitivity obeys the simple differential equation of Delbrück and Reichardt and only then in cases where a small stimulus was employed.

An interesting mathematical model was introduced by Lipson (1975 b) which defines the molecular nature of the level of adaptation. He assumes that the level of adaptation A is solely a function of the fraction of active photoreceptor. His model predicts that the time constants for adaptation and pigment regeneration should be identical. This prediction is consistent with the values he obtains by his white noise analysis of *Phycomyces* light-growth response system (Lipson, 1975a, 1975b).

c) Saturation of the Light-Growth Response

Foster and Lipson (1973) have shown that the response, R, to a pulse stimulus is saturated for $S \gg 10^2$ min. This is true also for step-up experiments when $I \gg 10^2 I_0$ (unpublished results). Saturation of this type occurs at low light intensities of the order of 10^{-3} W/m^2, and is therefore most unlikely to be due to bleaching of the photoreceptor. A subsequent biochemical step along the sensory transduction chain is most probably the limiting reaction. It is possible to ask if this limiting biochemical step is in the early part or the late part of the chain. (The late part of the chain is defined as the part common to both phototropic and autochemotropic responses.) The experiment consists of giving simultaneously two stimuli, a saturating step-up of light and a "house" (see Sect. E. 2.). Our results (unpublished) indicate that the house stimulus does not increase the response indicating that saturation occurs at a stage late in the sensory transduction chain. If the house stimulus is given 10 min

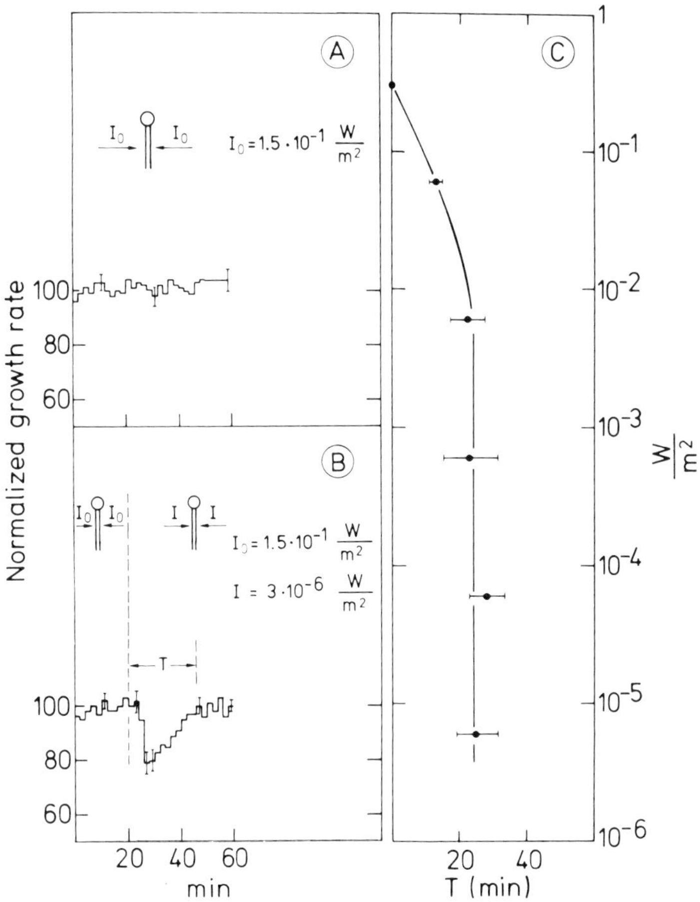

Fig. 7a. Normalized growth rate of the sporangiophore at constant light intensity I_0 = 1.5×10^{-1} Wm^{-2}. **b** Dark-growth response of the sporangiophore after a step-down of light (at 20 min). T is the time at which the growth rate returns to the baseline level. **c** Time T as a function of the light intensity after step-down of light. The values are obtained from a series of experiments of the type presented in Fig. 7b. The sporangiophores were always pretreated at I_0 = 1.5×10^{-1} Wm^{-2} and the step-down was made with different light intensities. The ordinate shows the total light intensity I after a step-down. All experiments were repeated on 4 different sporangiophores. The error of the mean is given. Sporangiophores were pretreated for 60 min at a total light intensity of 2 I_0 = 3×10^{-1} Wm^{-2} (unpublished results)

after the light stimulus then growth is stimulated again demonstrating that return of the growth rate to the base-line level after a light step-up (at about 10 min) is not due to an inability of the cell to continue growth at the faster rate as might occur if a vital metabolite were depleted.

d) Loss of Growth Rate Regulation

The growth response to a step-up of light does *not* adapt if the final light intensity is above 1 W/m^2 (Bergman et al., 1969; Foster and Lipson, 1973). Growth rate is maintained higher than the baseline growth rate. Foster and Lipson postulated that functional photoreceptors are inhibitors of growth and that under these conditions the majority of photoreceptors are converted into nonfunctional photoproducts and therefore the inhibitory control of growth rate is removed. We have confirmed their results for *wt* and a *madA* mutant but found that the breakdown of growth adaptation does not occur with *madB* mutants (unpublished results).

e) Sunrise Experiment

Bergman et al., (1969) have shown that it is possible to have a sustained growth response for 30 min if the light intensity is doubled every 2 min. This was the first indication that the return of the growth rate to the baseline level after a step-up cannot simply be due to a depletion of an important metabolite.

f) Light- and Dark-Growth Responses of the Phototropic Defective Mutants

Mutants in genes *madA* and *madB* behave as *wt* in light- and dark-growth responses at light intensities where they are phototropic. (The only exception to this result is discussed in Sect. E.3.d).

Mutant C148 *(madC)* has a normal light-growth response above 10^{-3} W/m^2 but does not have a dark-growth response (unpublished results). Mutants in the genes *madD, E, F, G* have a reduced amplitude in the light-growth response (Foster and Lipson, 1973).

4. Phototropism

a) Phototropism Under Constant Light Intensity

Phototropism under constant light intensity is the response to a unilateral stimulus having a light intensity identical to the total light intensity of the pretreatment light. The kinetic of the phototropic response can be very different if the light stimulus has an intensity different from the intensity of the pretreatment light. Measurable parameters of the phototropic response are the delay before bending starts and the rate of bending. The delay in phototropism is 6 ± 2 min and it is independent of the light intensity of the stimulus in a range from 1 to 10^{-6} W/m^2. The bending rate is of the order of 2 to 6 degrees/min and is also almost independent of the absolute light intensity of the stimulus in the same range of 1 to 10^{-6} W/m^2 (see Fig. 6). A noteworthy phenomenon is that the phototropic response can last for hours if the geometrical relationship between the growing zone and the light source is kept constant (Dennison, 1965). The growth response, however, lasts only 20 to 30 min. Several attempts have been made to explain this paradox (see Sect. E.5.). The bending rate ($d\alpha/dt$) is a function of the asymmetry of growth rate, ϵ, (due to an asymmetry of irradiation across the cell (focus advantage)) the cell radius, r, and average growth rate, $\bar{V}$:

$$\frac{d\alpha}{dt} = \epsilon \cdot \bar{V}/r$$

(Shropshire, 1971).

An intriguing and poorly understood phenomenon is the "aiming error". During a phototropic response the sporangiophore initially bends in a plane which is 45° to the direction of the light stimulus. A sporangiophore would bend towards an observer if the light were coming from the observer's right. After a further 15 to 20 min it would bend in a plane, parallel to the light stimulus and to the sporangiophore direction. By this time it would have deviated about 20 degrees from a plane perpendicular to the direction of the light stimulus (Dennison, 1965; Dennison and Foster, 1977; unpublished results). From the results of Dennison and Bozof (1973) where the tropic stimulus was ultraviolet light for only one minute, it can be concluded that the aiming error is probably due to the fact that the response is turning clockwise like the sporangium.

Normally in studies of phototropism the projection of the bending angle on a plane parallel to the light stimulus direction and to the sporangiophore is measured. The aiming error does not affect measurements of the delay of bending or of the bending rate as long as this is measured in the steady state part of the curve (about 20 min after the stimulus is given).

b) Phototropism After a Step-Down of Light Intensity

Delay in the phototropic response is larger than 6 min if the unilateral stimulus has a lower intensity than the pretreatment light. The delay as a function of the stimulus light intensity for two different pretreatment light intensities is shown in Fig. 8. The first point to note is that the delay is almost independent of the absolute light intensity of the pretreatment in a range of 10^4. It is, instead, a function of I/I_0. Furthermore the curves for *wt* are biphasic with a time constant of $b = 3$ min and $b = 9$ min. Interestingly mutants in gene *madA* behave almost like *wt* while mutants in gene *madB* seem to lack the fast part of the curve and adapt with a unique $b = 11$ min. No model has yet been proposed to explain how a mutant which is nightblind (by a factor 10^5) can also be defective in adaptation. The phenomenon of dark adaptation of phototropism could, in principle, be explained by a modified Delbrück and Reich-

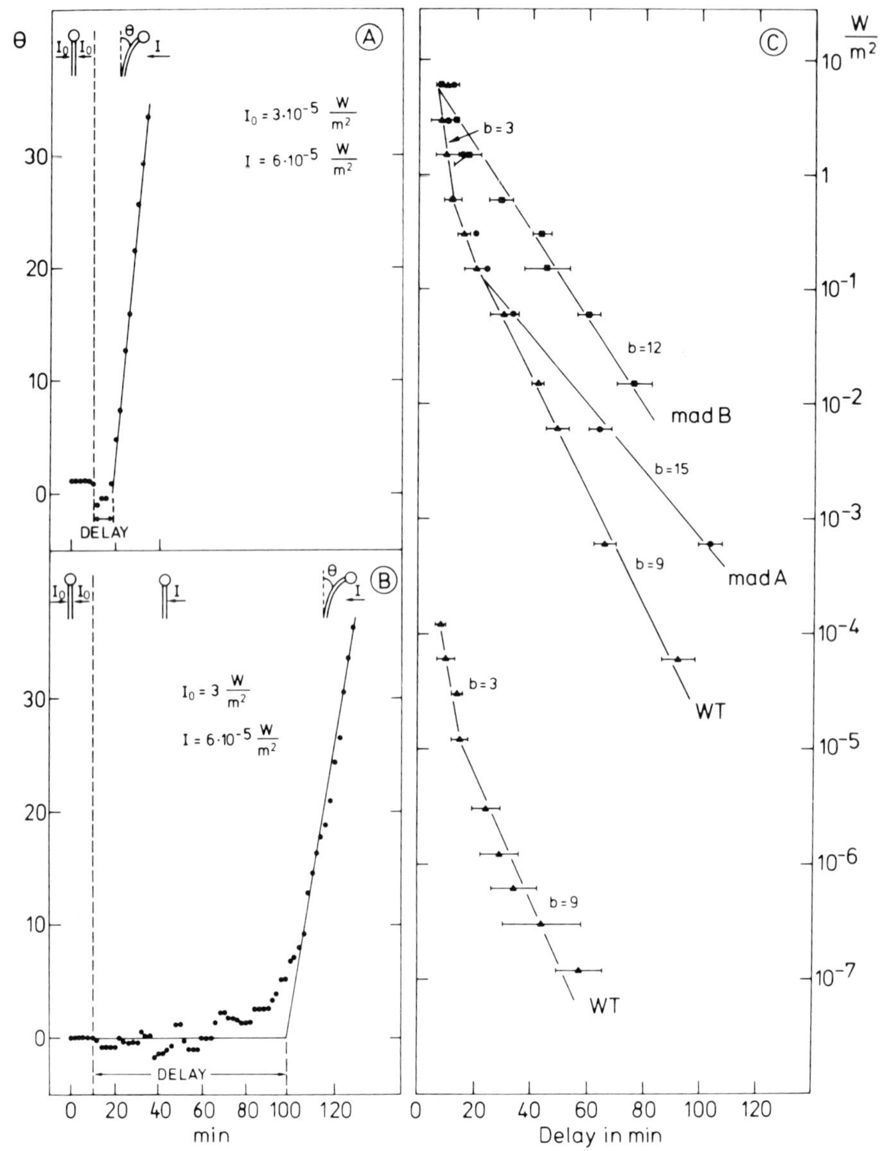
θ
A
$I_0 = 3 \cdot 10^{-5}$ W/m²
$I = 6 \cdot 10^{-5}$ W/m²
DELAY
B
$I_0 = 3$ W/m²
$I = 6 \cdot 10^{-5}$ W/m²
DELAY
min
C
W/m²
b = 3
b = 12
mad B
b = 15
b = 9
mad A
WT
b = 3
b = 9
WT
Delay in min

◁ **Fig. 8a.** Phototropic response at constant light intensity. The sporangiophores were pretreated for 120 min at a total light intensity of $2\,I_0 = 6 \times 10^{-5}\ Wm^{-2}$. At time 10 min the illumination was shifted from bilateral, to unilateral keeping the same total light intensity $I = 6 \times 10^{-5}\ Wm^{-2}$. The delay is obtained by extrapolating back from the steady state bending. **b** Phototropic response after a step-down of light intensity. The sporangiophores were pretreated for 60 min at a total bilateral illumination of $2\,I_0 = 6\ Wm^{-2}$. At time 10 min the illumination was made unilateral with an intensity of $6 \times 10^{-5}\ Wm^{-2}$. The delay is obtained in the same way as in Fig. 8a. **c** Delay of the phototropic response after a step-down of light intensity as a function of the light intensity of the phototropic stimulus I. In the upper part of the figure are given the curves for *wt*, *madA*, *madB*, all pretreated for 60 min at $2\,I_0 = 6\ Wm^{-2}$. In the lower part of the figure is a curve for *wt*, the pretreatment light intensity is $2\,I_0 = 1.2 \times 10^{-4}\ Wm^{-2}$. The points are the average of experiments on 4 different sporangiophores. The error of the mean is given (unpublished results)

---

ardt model except with the modification that the sporangiophore starts to bend when the level of adaptation A (or sensitivity) is equal to the phototropic light stimulus; experimentally two time constants instead of one are found.

c) Phototropism After a Step-Up of Light Intensity

In analogy to the experiments described above one obtains that the delay of a phototropic response will increase with increasing I/I_0. The delay as a function of the stimulatory light intensity is given in Fig. 9.

Following a step-up by a factor of 10^4 the sporangiophore does not bend for about 20 min. If a barrier is placed near the sporangiophore at the same time as the step-up of light then the autochemotropic response has a delay of 10 min. This indicates that the inability to sense a unilateral light stimulus is *not* in the late part of the sensory transduction chain. It is probable that a saturation occurs in the early part of the sensory transduction chain so that quantal changes in information are not transmitted for the first 20 min. The increased delay in the autochemotropic response (10 min instead of 5 min) probably results from saturation of the growth response during the first 10 min (see Sect. E.3.c).

d) Phototropic Indifference

At high light intensities sporangiophores do not bend (Fig. 6). This phenomenon can be explained by assuming that the photoreceptors on both the distal and the proximal sides of the sporangiophore are completely bleached. Alternatively a later biochemical step in the reaction chain may be saturated. In both cases the sporangiophore will be unable to sense a gradient of light and therefore does not bend. It has been suggested by Foster and Lipson (1973) that this phenomenon could be used to isolate specific mutants in the photoreceptor (see Sect. H.5.).

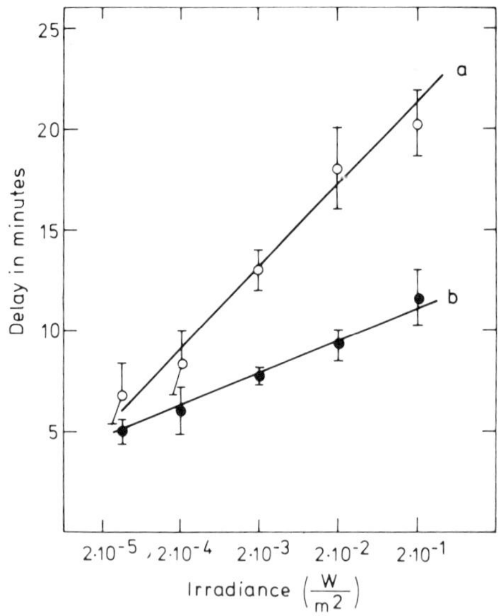

Fig. 9a. Delay of phototropism after a step-up of light versus the light intensity of the stimulus. The sporangiophores were pretreated at a total intensity $2 I_0 = 2 \times 10^{-5}$ Wm^{-2} for 90 min. The experiments are similar to the one of Fig. 8b with the difference that at the time of shift from bilateral to unilateral light the intensity of light stimulus was *increased.* **b** Delay of an autochemotropic response after a step-up of light. The sporangiophores were pretreated for 90 min at $2 I_0 = 2 \times 10^{-5}$ Wm^{-2}. At time 10 min the total light intensity was increased from 2×10^{-5} Wm^{-2} to the values written on the abscisse and a barrier was put in the vicinity of the sporangiophore. For this experiment it is irrelevant whether the illumination after the step-up in intensity is bilateral or unilateral (unpublished results)

e) Temporary Inversion of Tropic Responses

Reichardt and Varjú (1958) discovered that step-up in light intensity during unilateral irradiation causes a temporary inversion of the steady state phototropic reaction. Castle (1961) has demonstrated that a saturating pulse of light given from any direction can cause the inversion and Dennison (1965) has shown that a step-down will cause an inversion (Fig. 10). Removal of the light during a phototropic response also results in an inversion (unpublished results). At an intensity of 10^{-2} W/m^2 sporangiophores bend at a rate of 2.5 degrees/min; after the light is turned off reversed bending at a rate of 1.5 degrees/min takes place. Bergman et al., (1969) reported that an autochemotropic response or a geotropic response can also show an inversion if the sporangiophore is stimulated by a saturating pulse of light. Bergman et al., (1969) results therefore indicate that the difference in growth reactivity between distal and proximal sides can be attributed to a late part of the sensory transduction chain. Our light-off experiments suggest that a gradient of an "inhibitor" of growth may be inside the cell and that the distal side has more "inhibitor" than the proximal side.

f) Localization of Adaptation

To understand the molecular mechanism of phototropism it is essential to know whether adaptation is localized in a defined region of the cell or whether it is spread throughout the cell. To determine this Delbrück and Varjú (1961) performed a series of experiments in which they stimulated the sporangiophore with ultraviolet light in sharply bounded and narrow test areas (line width of 0.2 mm). They found that the

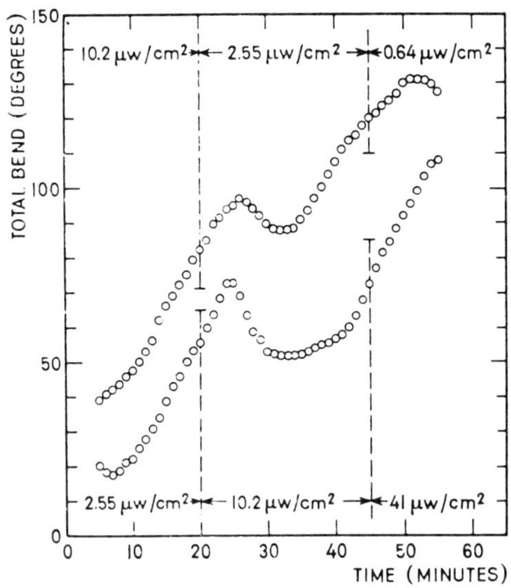

Fig. 10. Phototropic inversions following an increase in intensity *(lower curve)* and a decrease of intensity *(upper curve)*. (From Dennison, 1965)

adaptation (or sensitivity) does not spread along the sporangiophore axis. Furthermore they concluded "that the receiving and adapting structures do not move relative to the sporangium, and that the responding system does not move relative to ground" (Delbrück and Varjú, 1961). Dennison and Bozof (1973) went further in the study of this question and convincingly showed that adaptation also does not spread radially. Even more interesting, their experiments suggested that both the adaptation and the response are turning clockwise as does the sporangium itself (see Sect. B).

5. The Castle Model

All formal models of phototropism try to determine two main points: (i) is it possible to predict the phototropic response knowing the light-growth response? (ii) if so, how is it that in the light-growth response the growth rate returns to the baseline level after 20 min while the phototropic response is *constant* for hours?

Castle (1966) suggested the following scheme for the case of symmetrical irradiation

$$\xrightarrow{c} \mathrm{M} \xrightarrow{(k'+k)\,[\mathrm{M}]} \mathrm{P} \xrightarrow{j\,[\mathrm{P}]} \text{wall formation}$$

where c is the constant rate supply of M; [M] and [P] are the respective concentrations of these materials; k′ and k are the light-dependent and the light-independent rate constants of first reaction, and j is the rate constant of the second reaction. With this simple model he can explain qualitatively the various growth responses to the light programs of Fig. 5.

In the case of phototropism the scheme suggested is

$$\xrightarrow{c} M_2 \xrightarrow{(k'a + k)\,[M_2]} P_2 \xrightarrow{j\,[P_2]} \text{far wall formation}$$

$$\uparrow c'$$

$$\xrightarrow{c} M_1 \xrightarrow{(k' + k)\,[M_1]} P_1 \xrightarrow{j\,[P_1]} \text{near wall formation}$$

$$\uparrow \text{direction of light infall}$$

where a is a factor greater than one that denotes the photochemical advantage of the far half of the sporangiophore and c' expresses the lateral transport (or "leakage") of M that supports the consistently faster growth of the far side. One of the interesting points of the model is that $M_2 < M_1$. It is this characteristic that allowed the model to explain qualitatively all the inversion phenomena. With this model many of the known phenomena on light-induced growth and on phototropism and in particular the inversion phenomena, the steady phototropism, and the adaptation of growth could be explained qualitatively. There are doubts that it can explain the dark adaptation kinetics of growth (Fig. 7) and of phototropism (Fig. 8). The greatest drawback is that the model is only qualitative and not quantitative. No one has tried to fit the model to the actual data in order to determine the different parameters. For this reason one might question the argument of Bergman et al., (1969) that the sunrise experiment is in contradiction with the Castle model.

6. The Dennison-Foster Model

A model was presented by Dennison and Foster (1977) to explain the paradox that the phototropic response is steady while the growth response is limited. They suggested that the steady phototropic response is caused by the fact that the sporangium and the growing zone are in continuous clockwise rotation. This implies that the photoreceptor moves from an area where there is less light (the two flanks of the sporangium) to an area where there is more light (the back of the sporangiophores) due to the lens effect (see Fig. 2). This induces a continuous growth on the back side, namely a steady phototropic response. To support the model Dennison and Foster (1977) showed that the bending speed is diminished by a factor of two if the sporangiophore is rotated 10°/min counterclockwise. Their model does not, however, explain the steady phototropism of stage I sporangiophores in which the growing zone rotates at less than 20 degrees/h compared to the 720 degrees/h of Stage IV sporangiophore (Castle, 1936; unpublished results).

F. Photodifferentiation

In the rather long history of *Phycomyces* research emphasis has clearly been on the analysis of short term reactions, mainly light-growth response and phototropism, and analyses of photomorphoses have only been undertaken in recent years.

Three kinds of photomorphoses have been studied in *Phycomyces*: (a) light stimulated synthesis of carotene: photocarotenogenesis, (b) light induced formation of sporangiophores: photophorogenesis or photoinitiation, and (c) light induced formation of sporangia: photosporangiogenesis.

Although action spectra for these photoresponses have not been established the reactions were very helpful in establishing a formal ordering of the phototropic defective mutants (see Sect. D) and in demonstrating that the same photoreceptor of phototropism is also very likely active in the mycelium.

1. Photophorogenesis

a) Factors Influencing Photophorogenesis

Under standard growth conditions *Phycomyces* growth and differentiation is independent of light or dark treatments. Thornton (1972, 1973, 1975) found that under severely limited nitrogen supply *Phycomyces* can alternatively choose between two fruiting pathways; in the dark mainly dwarf sporangiophores will develop while illumination with a brief pulse or continuous blue light triggers formation of normal sporangiophores. An important factor in this system is low temperature since at 20 °C no pronounced light effect is observed.

Bergman (1972) took advantage of another system; his method essentially consists of using phosphate enriched potato dextrose agar in long "racing tubes". Little pieces of mycelia or spores are placed at one end of the tube and the growing mycelium is exposed each day to a short pulse of light. Under these conditions sporangiophore development is initiated the next day and is located at the point where the leading edge of the mycelium had been at the time of irradiation. Consecutive irradiations given at intervals of several hours results in a typical band pattern of sporangiophores reflecting the time program of irradiation. Light induced rhythms as found in *Pilobolus* (Uebelmesser, 1954) were not found in *Phycomyces*. Mutants, analogous to the *timex* strain found in *Neurospora* (Sargent and Briggs, 1967), which reveal spontaneous or light induced rhythms in sporangiophorogenesis in this organism have also not been isolated in *Phycomyces*. Red light is not effective in *Phycomyces* and the phototropic defective mutants *madA* and *madB* have a higher threshold in photophorogenesis. In contrast mutants in genes *madC* to *madG* exhibit normal photophorogenesis. This suggests that the photoreceptor involved in phototropism functions even in the mycelium.

A different type of photophorogenesis has been studied in our laboratory (Russo, 1977b). Here the formation of sporangiophores can be brought under blue light control by placing inoculated shell vials in tightly sealed glass beakers in the dark.

Under these conditions the growing mycelia change the atmosphere in such a way that growth of sporangiophores can be completely repressed. The degree of repression depends on the amount of mycelia per beaker, e.g., with 70 vials/1 l beaker almost no phorogenesis will take place. This repression is lifted if the mycelia are irradiated with a short pulse of blue light. Various aspects of this phenomenon have been analyzed to determine the nature of the inhibiting gas(es), their regulation by light and the elaboration of fluence response curves.

b) Fluence Response Curves

The test system of photophorogenesis developed in our laboratory has the advantage of greater precision than the system previously described by Bergman (1972). This has permitted a critical reexamination of *madA* and *madB* mutants and various carotene lacking mutants for light sensitivity (Galland and Russo, 1979a, in press). Figure 11 shows that the threshold for photophorogenesis for the wild type, NRRL-1555, is about 6×10^{-5} W/m^2 (1 min irradiation). The phototropic defective *madA* and *madB* mutants have a raised threshold whereas mutants *madC* and *madD* to *madG* are not affected. This result confirms the findings of Bergman (1972) despite the differences in the test systems. Surprisingly all mutants lacking β-carotene were found to have a raised threshold directly correlated to the residual amount of β-carotene (Fig. 12). This result was not anticipated particularly as these mutants exhibit normal phototropism (Presti et al., 1977). It appears therefore that β-carotene

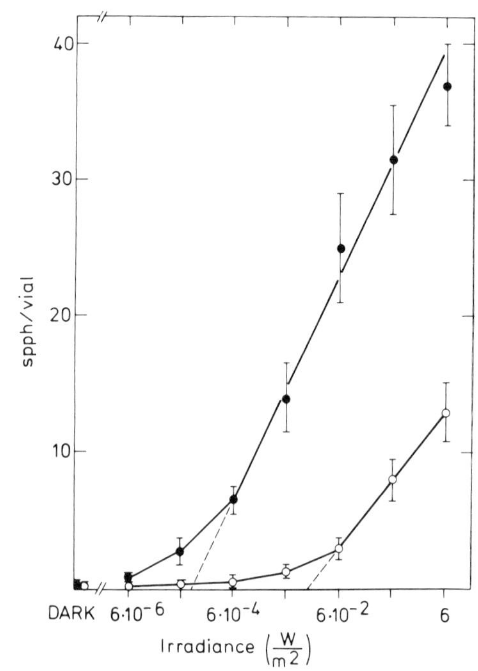

Fig. 11. Fluence response curves of photophorogenesis. Mycelia, 50 h after inoculation, were irradiated for 1 min at the indicated intensities. Filled circles = *wt* NRRL1555; open circles C173 *(car B32 car R21)*. From Galland and Russo (1979 in press)

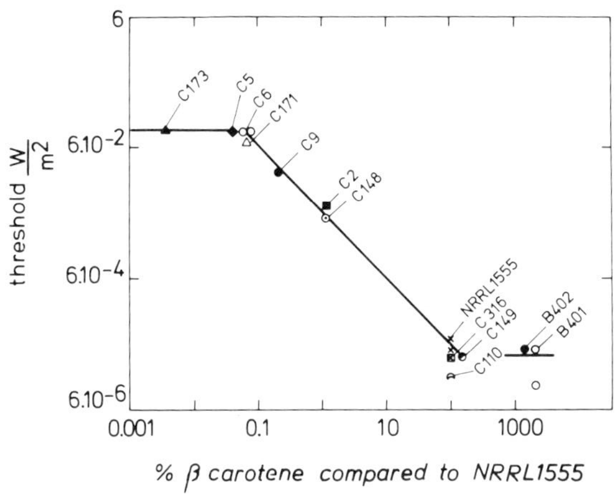

Fig. 12. Light threshold of photophorogenesis in dependence of residual β-carotene synthesis. Mycelia, 50 h after inoculation, were irradiated at various fluence rates for 1 min. The threshold was determined by extrapolation. From Galland and Russo (1979, in press)

does not play a direct role in light perception resulting in phototropism but functions in the light pathway operating in photophorogenesis. Evidence currently available suggests that a flavin acts as the photoreceptor in photophorogenesis and therefore β-carotene (or a related metabolite) must act as a "messenger" molecule functioning at a stage later than the photoreceptor (Galland and Russo, 1979a, in press).

c) The Role of Retinol

Analysis of phorogenesis inhibition observed in the closed system has shown that oxygen and retinol play vital roles in differentiation. These compounds interact in a complex way with one another and with light (Galland and Russo, 1979b, in press).

Lack of oxygen inhibits the formation of sporangiophores in a closed system. Caroteneless mutants are more sensitive to the atmosphere in the beaker than is the wild type. Reconstitution experiments, with varying amounts of oxygen, have shown that caroteneless mutants have a much higher oxygen requirement for phorogenesis than has the wild type. Carbon dioxide does not affect phorogenesis. A short irradiation of the mycelium with blue light leads to full induction of sporangiophores at limiting oxygen concentrations which completely inhibit phorogenesis in the dark.

Retinol has been known for many years as an important regulator molecule in the growth and differentiation of animal cells (Moore, 1967). There is good evidence that retinol induces the synthesis of specific proteins and *mRNAs* (Blalock and Gifford, 1977; De Luca et al., 1971; Kaufman et al., 1972). Retinol induced differentiation in plants or fungi is, as yet, unknown and, indeed, the biosynthesis of retinol in plants and fungi has not yet been unambiguously demonstrated. The role of retinol in *Phycomyces* phorogenesis was discovered when it was found that the extremely high oxygen threshold for phorogenesis exhibited by the caroteneless mutants can be suppressed by retinal (Fig. 13), retinol or retinolacetate (Galland and Russo, 1979b, in press). Retinol appears to be a rather specific requirement for phorogenesis as the caroteneless mutants show no defects in respiration, oxygen requirement for myceliar growth, or sporangiogenesis.

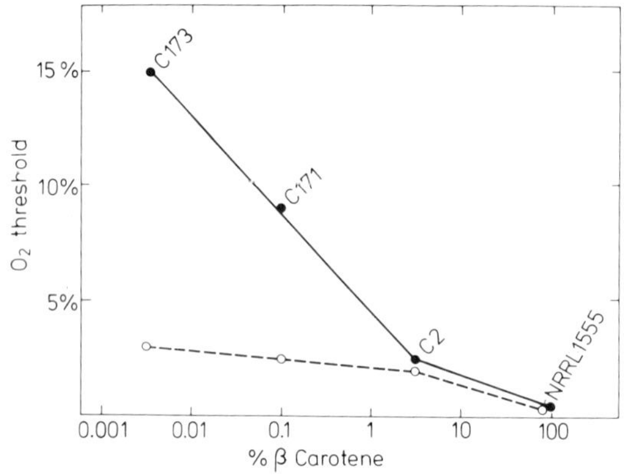

Fig. 13. Oxygen threshold of phorogenesis in dependence of residual β-carotene synthesis as compared to *wt* NRRL1555 (= 100%). ●– – –● = no addition of retinal; ○– – –○ = addition of 125 μg/ml retinal to the medium. From Galland and Russo (1979 in press)

2. Photosporangiogenesis

Formation of sporangia like phorogenesis is rather independent of light or dark. Conditions have, however, been found which bring sporangiogenesis under blue light control. Sporangiogenesis is delayed for many hours if stage I sporangiophores are kept in tightly sealed beakers. Irradiation, with continuous blue light reverses the inhibition which is caused by the accumulation of a gas (unpublished results). Experiments to date suggest that oxygen is not a limiting factor under the conditions used; carbon dioxide and hydrogen cyanide stimulate sporangiogenesis and ethylene and β-ionone partially inhibit sporangiogenesis.

Continuous light has a very complex effect on sporangiogenesis in the closed system as can be seen from Fig. 14. Synchronized stage I sporangiophores (Russo, 1977b), enclosed in 1 l glass beakers (time 0), were kept at different light intensities. After 2 days the percentage of differentiated sporangiophores, i.e., those containing a sporangium were plotted as a function of light intensity. Light of low intensities exerts an inhibitory effect and light of high intensities a stimulatory effect. Phototropic defective *madA* and *madB* mutants have raised thresholds for both light reactions, i.e., the inhibitory and the stimulating effects. The threshold for the inhibitory reaction is raised by a factor of 10^7 whereas the threshold for the positive part is raised by only a factor of 10^2. At the moment the bizarre intensity dependence curve for photosporangiogenesis cannot be explained.

3. Photocarotenogenesis

Light induced accumulation of β-carotene is one of the most remarkable features of *Phycomyces* and is obvious to anyone working with this organism. More than 97% of the total carotenoids exist as β-carotene (Hsu et al., 1974) which makes it especially easy to follow gross changes in the pigment after light induction. Photocarotenogenesis is for several reasons an interesting system to study. It offers itself as a model system to study a light induced change in a biosynthetic pathway as the endproduct of

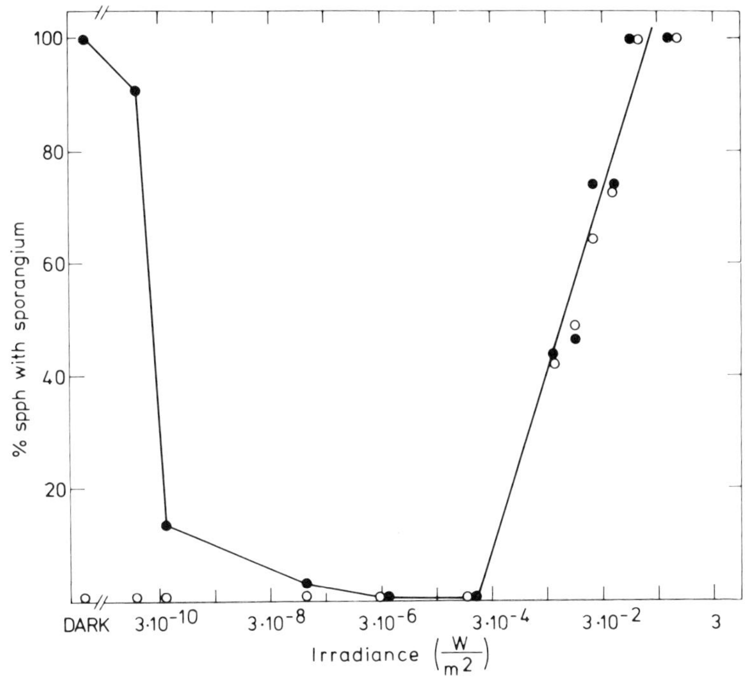

Fig. 14. Photosporangiogenesis of *wt* NRRL1555. Stage I sporangiophores were kept continuously under blue light. *Open circles* = nr. of sporangia after 24 h of illumination; *filled circles* = nr. of sporangia after 48 h of illumination. Russo and Galland (unpublished)

the pathway is optically very easy to detect. It would be interesting to determine if light acts at the level of transcription or at the level of translation. This approach has already been successfully applied to *Fusarium aquaeductum* (Rau, 1975; Schrott and Rau, 1977) and has led to the concept of transcriptional regulation of carotenogenesis by light. Although there is much evidence which argues against β-carotene as the blue light photoreceptor, it is clear that β-carotene plays a role in photophorogenesis (see Sect. F.1.b) and it is therefore of vital importance to come to an understanding of how the synthesis of this molecule, itself, is controlled by light. Despite these various aspects the systematic study of photocarotenogenesis has only been taken up in recent years. Early work on carotenogenesis in *Phycomyces* was undertaken with the main intention of elucidating the biosynthetic pathway leading to β-carotene and light induction of this process was not a research problem of major concern. Light stimulated carotenogenesis was, however, initially reported by biochemically oriented research groups (Garton et al., 1951; Goodwin et al., 1956). Under the conditions applied by these authors only a 2-fold stimulation of carotene by light was found. In 1973 Bergman et al., published a dose response curve which showed that under continuous illumination (10^{-1} W/m^2) the amount of β-carotene increases about 20-fold. Light induced carotene synthesis starts about 1 hour after initiation of continuous white light illumination (Sandmann and Hilgenberg, 1978). This light reaction requires the wild type allele of the *madB* gene but mutations in genes *madD* to *madG* do not affect photocarotenogenesis (Bergman et al., 1973). A most interesting result came from analyses of *carA* mutants which have a defect in the synthesis of β-carotene. These mutants have a residual level of β-carotene of about 1% to 3% of the *wt* amount. One might expect that the residual carotene synthesis exhibited by these mutants would be stimulated by light. This is, however, not the case; a light intensity of 2.1 W/m^2 which stimulates carotene synthesis in *wt* about tenfold, does not stimulate β-carotene synthesis in the mutants (Eslava et al., 1974). Mutants in the *carB* and *carR* genes accumulate phytoene and lycopene but light does not increase the amount of the compounds accumulated (Eslava and Cerdá-Olmedo, 1974). Thus, one is led to the conclusion that β-carotene, or a metabolite of β-carotene, is itself a prerequisite for photocarotenogenesis. Indeed, the behavior of the mutants in photocarotenogenesis is very much reminiscent of their behavior in photophorogenesis in which they demonstrated a higher visual threshold. These observations alone do not, however, permit any conclusion as to whether or not β-carotene functions as photoreceptor.

The regulation of β-carotene synthesis is remarkably complex and is still far from being elucidated. Isolation of mutants blocked at various steps along the biosynthetical pathway has led to some progress in this respect (Ootaki et al., 1973; Meissner and Delbrück, 1968). The analysis of various mutants supports the idea that the biosynthetic route is regulated by end product inhibition, i.e., β-carotene blocks its own synthesis. This conclusion is based on the observation that in mutants *carB* (accumulation of phytoene) and mutants *carR* (accumulation of lycopene) the accumulated precursors are present in much larger quantities than the total amount of carotenes in the wild type (Eslava et al., 1974). End product inhibition as a control mechanism is also supported by the observation that β-carotene inhibits its own synthesis in cell-free extracts of *Phycomyces* which are capable of converting mevalonic acid into isoprenoids and β-carotene (Bramley and Davies, 1976). One remarkable and potentially very interesting property of β-carotene biosynthesis is the stimulation by either β-ionone (Mackinney et al., 1952; Reyes et al., 1964) or retinol (Eslava et al., 1974). The mechanism of this stimulation remains unclear, however, it is known that neither of these compounds is a precursor of β-carotene. β-Ionone stimulates general isoprenoid synthesis (Reyes et al., 1964) and retinol acts at a stage prior to phytoene synthesis (Eslava et al., 1974).

G. Biochemical Aspects of the Light Transduction Pathway

Little is known at present about the biochemical reactions involved in light promoted growth stimulation.

Cyclic nucleotides which are involved in vertebrate vision (Miki et al., 1973) also seem to play a role in the light-growth response of *Phycomyces*. Cohen (1974b) showed that 1 min after a step-up of light the level of cAMP of Stage IVb sporangiophores drops transiently by about 60%. The drop in the cAMP pool does not appear to be a spurious by-product of the light intensity change but a real constituent of the light transduction pathway as cAMP given exogenously to sporangiophores elicits a transient decrease in growth rate. Recently it was shown by Cohen and Atkinson (1978) that blue, but not red light, stimulates cyclic AMP phosphodiesterase extracted from sporangiophores. The enzyme activity is enhanced by 1–10 μM FAD and FMN and 1 mM GTP. That cyclic nucleotides are involved in the light growth regulation was convincingly shown in a recent report of Cohen (1978). The transient decrease of cAMP in sporangiophores after a step-up of light is about sixfold smaller in the mutant C 149 *(madD 120)*. This mutant belongs to the class II mutants (see Sect. D), which have the same phototropic threshold but a largely reduced bending rate and a reduced light-growth response. Interestingly the pH sensitivity of the phosphodiesterase is altered in this mutant. Adenylate cyclase has been demonstrated histochemically (Tu and Malhotra, 1973) and is present in the plasmamembrane, nuclear membrane and mitochondrial membranes in all stages of *Phycomyces* development.

Light must exert an effect on the growth of the cell wall. Approximately 40% of the cell wall (dry weight) consists of chitin [poly-β(1–4)N-acetyl-D-glucosamine] and its nonacetylated form (Cohen, 1974a) and chitin synthetase activity of *Phycomyces* sporangiophores was found to be approximately 30% stimulated by light (Jan, 1978). In yeast proteinases convert the inactive chitin synthetase zymogen into an active enzyme (Ulane and Cabib, 1976). A similar mechanism seems to exist in *Phycomyces*. Recently Fischer and Thomson (1979, in press) characterized three serine proteinases of *Phycomyces* which are able to stimulate chitin synthetase activity. All of these serine proteinases could complex with one of two proteinaseous inhibitors. Fischer and Thomson also isolated an acid proteinase from *Phycomyces* which removes the inhibitors from the serine proteinase – inhibitor complex. Crude extracts of three of the class II mutants displayed differences in proteolytic activity and a fivefold increased acid proteinase activity in comparison to wild type; however, it is not yet clear if these changes are directly related to the inability of these mutants to bend.

Another enzyme involved in cell wall growth is chitinase which could be involved in the regulation of growth rate. Its properties in *Phycomyces* sporangiophores have been described (Cohen, 1974a) but light effects on the activity of its enzyme have not yet been reported.

H. The Blue Light Photoreceptor

To date, analyses of the *Phycomyces* light responses have remained at a rather formal level whereas the ultimate goal must be a detailed description of the chemical steps in reception, transmission and reaction to the stimulus of light. The first of these steps is excitation of the photoreceptor by light and it is therefore not unexpected that many attemps have been made to identify and isolate this compound. In spite of much effort during the past years, not only with *Phycomyces*, the blue light photoreceptor has remained elusive and for this reason information about its chemical nature must be determined by indirect means. Before a biochemical isolation of the photoreceptor becomes feasible, one therefore depends on and is in urgent need of chemical, physical, and genetical criteria by which the photoreceptor can be traced and hopefully identified. A prerequisite to the successful isolation of the photoreceptor is the absorption spectrum; information about its extinction coefficient, regeneration constant, and quantum efficiency will be helpful; such biophysical studies should be matched by a genetical approach to the problem. The availability of mutants which in molecular biology has been such a useful method in the process of protein identification, would be a powerful means to trace the photoreceptor and physiological criteria for such mutants must be elaborated.

1. Action Spectra

The best information available regarding the absorption spectrum of the blue light photoreceptor is from correlations with action spectra. The spectral sensitivities of the light growth response and phototropism were studied very early (Castle, 1931; Castle, 1935) and show properties similar to action spectra found in other lower and higher plants (see article of W.Schmidt in this volume). The precision by which action spectra can be determined is limited by screening caused by the visual pigment itself or by other pigments. This results in losses of light intensity and concomitant "dips" or "shoulders" in the spectra. Knowledge of these factors and localization of these pigments relative to the photoreceptor are therefore prerequisites for a valid estimate of the gross errors inherent in action spectra. It has been pointed out by Delbrück and Shropshire (1960) that *Phycomyces* offers an unique technical advantage in this respect as screening in the light growth and phototropic responses has opposite effects. Screening in the light-growth response causes a gross reduction of the actinic light and results in an apparent diminition of the relative quantum efficiency; on the other hand, in phototropism, the light gradient between distal and proximal sides becomes steeper causing an apparent higher quantum efficiency. Comparison of the two action spectra therefore allows an estimate of screening. A further advantage of the *Phycomyces* system is in the applicability of the null determination by which a test light, with a defined wavelength, has to neutralize the effect of a standard light source causing a standard response. In this way errors caused by variations of the sensitivity in individual specimens can be minimized. In the light-growth response a periodic stimulus of 5 min duration with a standard blue light source alternating with a light pulse of the same length with a test light of defined wavelength; the intensity of the test pulse must be monitored such that the sporangiophore does sense the two light sources as equivalent in intensity being recognized by absence of alteration of growth rate. An analogous procedure is applied in the case of the phototropic action spectrum; in this case one uses two opposed light sources one giving the standard and one giving the test stimulus; absence of bending toward either of the light sources indicates the equivalent energy. This method was applied by Curry and Gruen (1959) and Delbrück and Shropshire (1960). The action spectra reported by these authors agree reasonably well and show maxima at 280 nm, 385, 455 and 485 nm (Fig. 15). The

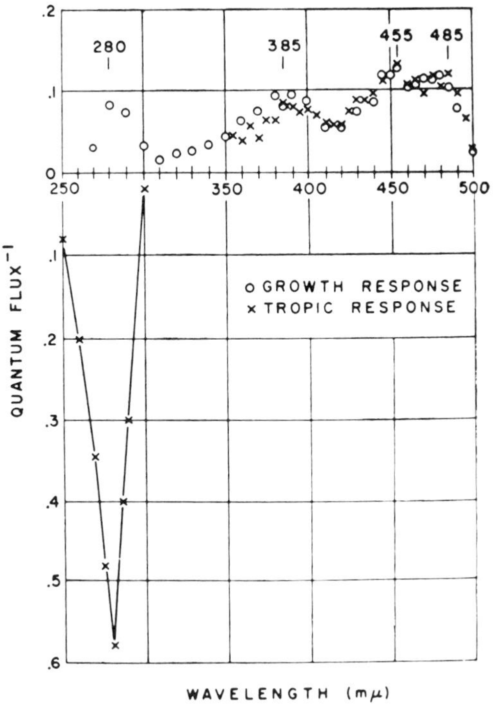

Fig. 15. Action spectra for growth response and tropic response. (From Delbrück and Shropshire, 1960)

peaks at 280 and 385 nm are indicative of a flavin and indicate that β-carotene is a rather unlikely candidate for photoreceptor. The action spectra of the light-growth reaction and phototropism are virtually identical in the visible range and from this it is concluded that the internal screen can be neglected and that the action spectra reliably reflect the absorption spectrum of the visual pigment (Delbrück and Shropshire, 1960). Below 300 nm phototropism becomes negative, i.e., the sporangiophore bends away from the light source and it can be shown that this comes from the internal screening of gallic acid (Dennison, 1959). The aforementioned action spectra have been worked out at intermediate intensities whereas action spectra can be established in which the absolute threshold of phototropism at different wavelengths is determined. In this method the null determination can naturally not be applied. An action spectrum determined in this way shows gross similarities to the spectrum obtained with the null method with respect to it having a sharp cutoff at 500 nm and 300 nm and a phototropic inversion below 300 nm. The pronounced peaks at 370, 450 and 485 nm are, however, missing (Varjú et al., 1961).

Delbrück et al., (1976) have recently sought to determine whether or not the stimulation of *Phycomyces* involves the excitation of the lowest triplet state of riboflavin. They determined an action spectrum of light-growth response between 575 and 630 nm using a tunable laser beam and taking advantage of the null method described above. This action spectrum was compared with an action spectrum obtained by computer extrapolation of a phototropic action spectrum covering 445–560 nm.

Comparison of the computed and the measured relative cross sections between 575 and 630 nm shows a difference which is indicative for the lowest triplet state of a riboflavin. The action spectrum has a small peak at 595 nm and a cutoff at 585 nm and a tail extending beyond 630 nm. The peak at 595 nm may be regarded, within the accuracy of this system, as the lowest triplet state of riboflavin and can be compared to the value of 600 nm obtained by phosphorescence emission for this state (Song and Moore, 1974).

2. *Light Induced Absorbance Changes*

A different, more direct approach to characterize the photoreceptor is based on the assumption that if a simple biochemical test were available to assay the presence in vivo and in vitro of the receptor pigment then the photoreceptor could be successfully isolated. Such a test could be a light stimulus causing bleaching or absorbance changes with regeneration constants comparable to the constants found in vivo. If such a light effect were available which also matched the spectral sensitivity of the in vivo action spectra, then one would have a way to further trace and purify the receptor pigment. Furthermore, one could expect differences in bleaching and absorbance changes in the phototropic defective mutants which have higher visual thresholds. This would provide an additional criterium that the observed effects are related to the physiological reactions. Research carried out with various organisms along this line is discussed in this book and it is still much debated to what extent the observed absorbance changes reflect integral parts of the sensorial transduction chain or whether they constitute spurious by-products of light excitation. To the best of our knowledge the first authors taking up this type of approach were Berns and Vaughn (1970) who looked for blue light induced absorbance changes in the wild type, a β-carotene lacking mutant and the phototropic deficient mutant C59 which is defective in the early gene, *madC*. Irradiation of intact or extracts of mycelia and sporangiophores resulted in an increase in ΔOD (OD 460 nm – OD 345 nm) in all three strains. Sporangiophores from the wild type and β-carotene lacking mutant achieved regeneration in 10–13 min whereas no regeneration occurred in the mutant C59 *(madC)*. Analysis of fluorescence excitation and emission spectra made with extracts of sporangiophores indicated bleaching of the wild type and *car* mutant but negligible bleaching of the *madC* mutant. Regeneration was complete after 5 min for *wt* and *car*$^-$ but, as stated above, the *madC* mutant showed no regeneration. The fluorescence emission of the *madC* mutant is reduced by a factor of two at 440 nm and a factor of three at 540 nm as compared to the *wt*. The fluorescence excitation and emission spectra indicate the presence of two pigments which are probably oxidation products of thiamine and a flavin (Berns and Vaughn, 1970).

Even though the reproducibility of the measurements of the absorbance changes have been questioned (Poff and Butler, 1974) we think that these results, if valid, are very promising. In particular they demonstrate that this light reaction is related to the in vivo sensorial transduction since the blind mutant C59 reveals quantitative differences in the fluorescence emission and excitation spectra and in regeneration. It may be very significant that the absence of regeneration correlates well with the observation that *madC* mutants have an extremely slow dark adaptation (Foster and Lipson, 1973). It is tempting to speculate that the lack of regeneration observed in vitro is related to the phenomenon of adaptation in vivo.

The majority of work on light induced absorbance changes has been undertaken with *Dictyostelium* and *Neurospora*. Absorbance changes in these organisms are triggered in vivo and in vitro by blue light which is indicative of a reduction of a b type cytochrome (Poff et al., 1973; Poff and Butler, 1974; Munoz and Butler, 1975). Similar experiments have been performed with mycelia of *Phycomyces* wild type and various phototropic defective strains (*madA*, *B*, *C*, *D*) with the intention of characterizing the mutants by possible differences in absorbance changes (Lipson and Presti, 1977). Blue light induced absorbance changes have indeed been found indicating a reduction of b and c type cytochromes however no differences between the mutants and *wt* strains were detected. The action spectrum of these light induced absorbance changes is suggestive a flavin photoreceptor. The authors themselves regard these effects as not necessarily being linked to the sensorial transduction event because the quantum yield which can be computed for this light reaction is extremely low (0.015) in comparison to the quantum yield computed for the light-growth reaction (~ 1.0) (Lipson, 1975a). In addition the same type of absorbance changes observed in *Phycomyces* is also observed in *Hela* cells which are cancer cells of the human brain which would not be expected to contain a physiological light transduction mechanism. This suggests that the whole phenomenon may be a spurious light effect rather than an integral part of the physiological light transduction.

3. Extinction Coefficient of the Photoreceptor

In addition to the action spectra regeneration constant and partial extinction coefficient of the photoreceptor have also been deduced by indirect means. Application of the technique of white noise analysis of the light-growth response demonstrated a reduced *Kernel* amplitude at high light intensities (Lipson, 1975a, b). This observation was interpreted in a model of inactivation and regeneration of the visual pigment with a regeneration constant of 2.7 min and a partial extinction coefficient $\epsilon^{I}_{455} = 1.5 \pm 0.2 \times 10^4$ liter·mol^{-1}·cm^{-1}. (The partial extinction coefficient is the total extinction coefficient multiplied by the quantum yield for inactivation.) Both of these values are comparable with the respective parameters for the human visual pigment (Alpern, 1971; Rushton and Henry, 1968). The closeness of ϵ^{I}_{455} to the extinction coefficient of oxidized flavin ($\epsilon_{446} = 1.25 \times 10^4$ l·mol^{-1}·cm^{-1}) adds further weight to the idea of a flavin being the photoreceptor of *Phycomyces*.

4. Flavin or β-Carotene?

A final criterium for the photoreceptor would be its presence at an adequate concentration. On the basis of a few reasonable assumptions Delbrück estimated the minimal theoretical concentration for photoreceptor to be about 3×10^{-7} M corresponding to 10^9 receptor molecules per growing zone (in Bergman et al., 1969).

This admittedly rough estimate has, however, served as a guide for several experiments devised to rule out β-carotene as the photoreceptor. The rationale behind these experiments was to determine if the lack of β-carotene caused by mutation (in *Phycomyces* and *Neurospora*) or by diphenylamine (in *Pilobulus*) affected blue light sensitivity (Meissner and Delbrück, 1968; Sargent and Briggs, 1967; Page and Curry, 1966). In none of these cases was the phototropic sensitivity reduced, however, this does not necessarily exclude β-carotene as a possible photoreceptor since in all cases the residual concentration of β-carotene was much higher than the critical concentration which was calculated to be required, i.e., 10^9 receptor molecules per growing zone. The availability of double mutants of *Phycomyces* which are defective in the four dehydrogenation steps and the two cyclisation steps leading from phytoene to β-carotene have allowed this obstacle of residual synthesis to be overcome since the residual amount of β-carotene in these double mutants is less than 4×10^{-5} of the wild type amount. This means that they contain less than the calculated critical concentration of the photoreceptor. The phototropical threshold of these double mutants is not altered (Presti et al., 1977) which strongly corroborates the conclusion drawn from the action spectra that β-carotene is very unlikely to be the visual pigment of phototropism in *Phycomyces*.

The normal phototropic response of β-caroteneless mutants with no trace amounts of carotene (Presti et al., 1977) is in apparent contrast to the deficiency exhibited by the same mutants in photophorogenesis. A clear relationship between β-carotene concentration and visual threshold for photophorogenesis could be established. However arguments have been presented indicating the unlikelihood that β-carotene is the only photoreceptor in photophorogenesis (Galland and Russo, 1979a, in press).

5. Physiological Criteria for Photoreceptor Mutants

Experiments with the β-caroteneless mutants were very valuable and have essentially excluded β-carotene as a candidate for the photoreceptor however it would be desirable to have positive physiological criteria for a mutant which is altered in the photoreceptor. Such criteria have been developed for photoreceptor mutants with respect to their behavior at low and high light intensities (Foster and Lipson, 1973). Phototropism of the wild type is inhibited at high light intensities (6–10 W/m², Fig. 6) which is explained as resulting from light saturation of the photoreceptor molecules at the distal and proximal site of the sporangiophore. Three types of photoreceptor mutants can be expected

(i) those which contain less photoreceptor than the wild type
(ii) those with the same photoreceptor concentration as wild type but with a smaller absorption cross section
(iii) those with the same photoreceptor concentration as wild type but with a slow regeneration after bleaching.

For all types of mutants one can make predictions for their phototropical behavior at low and high light intensities. These predictions are based on the following rationale:

it is assumed that the response R is a function of the number of bleached photoreceptor P_{bl}, the cross section σ and the light intensity I:

$$R = f(P_{bl}) = f(P_{tot} \cdot \sigma \cdot I) . \qquad (1)$$

Equation (1) predicts that at low light intensities mutants of type (i) or (ii) have a raised threshold, since either reduction of σ or P_{tot} can be compensated by raising I (Fig. 16). At high light intensities unbleached photoreceptor becomes limiting and one has to take into consideration the ratio P_{bl}/P_{tot}; this ratio depends on the intensity I and the cross section σ:

$$P_{bl}/P_{tot} = f(\sigma \cdot I) . \qquad (2)$$

At high intensities this ratio becomes unity (all photoreceptor is bleached) and therefore no phototropism can occur. Equation (2) shows that the bleaching of the photoreceptor does not depend on the absolute concentration of photoreceptor. Therefore mutants with less receptor pigment will be inhibited at the same light intensity as the wild type. However, in mutants with a reduced cross section the ratio of P_{bl}/P_{tot} will reach unity at higher intensities. Hence they are phototropic active at intensities where the wild type is already inhibited. The third type of mutant (iii) which has a slower regeneration as the wild type should have a normal sensitivity at low intensities but should be phototropical insensitive at higher intensities at which the unbleached

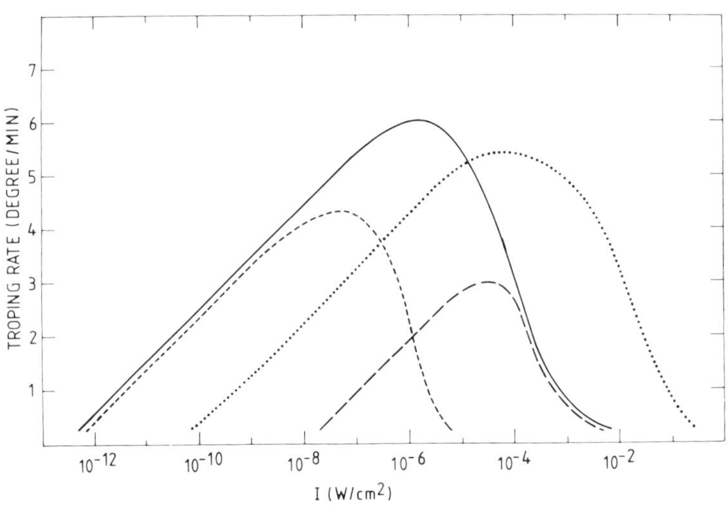

Fig. 16. Phototropic threshold of wild type and hypothetical photoreceptor mutants. *Solid line* = wild type; *large dashes* = mutant with reduced number of photoreceptor; *dotted line* = mutant with reduced absorption cross-section; *small dashes* = mutant with slow regeneration. The changes of threshold of the hypothetical mutants were chosen arbitrarily. The figure was adapted to Fig. 6 and the solid line of the wild type represent data of Foster and Lipson (1973)

photoreceptor becomes limiting (Fig. 16). This method is based on differential light absorption in the sporangiophore which causes the phototropism. This is an unique advantage of the structure and optical properties of the *Phycomyces* sporangiophore and therefore this method could not be transferred to other organisms which do not have analogous geometrical properties. An attempt to isolate mutants of these types in *Phycomyces* is currently being made (Lipson, personal communication).

6. Localization and Orientation of the Photoreceptor

The necessity of knowing the precise location of the receptor is self-evident with respect to a future isolation of the visual pigment and to an understanding of the mechanism of the light-growth reaction and phototropism. On the basis of a comparison of the action spectra and transmission spectra it was concluded that the photoreceptor must be localized between plasmalemma and tonoplast (Delbrück and Shropshire, 1960). This conclusion has been strongly supported by experimental work of Meistrich et al., (1970). These authors used a laser beam which was very small in comparison to the cross section of the sporangiophore to determine the position of maximal phototropical effect. The beam was allowed to horizontally traverse one half of the growing zone. A plasma membrane bound or cell wall bound receptor would be expected to exhibit maximal stimulation with a beam tangentially touching the sporangiophore but maximal tropic stimulation was found when the beam was focused on a position between cell wall and vacuole. The conclusion must be that the photoreceptor lies inside the cytoplasm.

The orientation of the photoreceptor can be studied by analyzing the effects of polarized light. Castle (1934) first discovered that when sporangiophores are symmetrically illuminated with longitudinally and transversely polarized light that the longitudinal beam has to be about 15% more intense than the transverse beam to avoid phototropic bending. Castle interpreted this result in terms of differential reflection losses rather than by dichroism of oriented receptor molecules. Castle's interpretation was apparently confirmed by Shropshire (1959) who showed that the higher effectiveness of transversally polarized light disappeard when the sporangiophore was immersed in a medium of the same refractive index as the sporangiophore, which diminished refraction losses. The interpretation of these results was subsequently challenged by Jaffe (1960) who calculated that reflection could only explain a minimal part of the differential effectiveness of polarized light beams. Jaffe's theoretical analysis predicts that reflection losses at the surface are compensated by internal reflections. Internal reflections in turn are attenuated by absorption mainly due to β-carotene. If the polarized light effect were only caused by reflection losses then one should expect a smaller effect of polarized light in β-carotene lacking mutants. Jesaitis (1974), however, could clearly show that the effect of polarized light is as strong in β-caroteneless mutants as in the wild type. Hence he could exclude that the polarized light effect comes from reflection losses; instead the effect of polarized light is due to linearly dichroic and oriented photopigment. The orientation of the dipols found in UV at 280 nm and in the blue range are compatible with the assumption of the flavin nature of the receptor as can be concluded from fluorescence polarization data (Kurtin and Song, 1968; Siodmiak and Frackowiak, 1972). A positive identification of the receptor in this way is not, however, yet possible as similar data for β-carotene are not available.

I. Summary

Among the many sensory reactions *Phycomyces* displays, the study of the photoreceptor and adaptation deserves maximal attention, since *Phycomyces* shares these two attributes with a variety of other blue light sensitive organisms. Action-spectroscopy indicates a flavin as the photoreceptor of *Phycomyces*. β-carotene was positively ruled out as a possible receptor, since mutants with no trace amounts of β-carotene are phototropical normal. The photoreceptor has not yet been isolated. As in other systems the difficulty consists in distinguishing the flavin photoreceptor from the bulk flavoproteins in the cell. One therefore needs unambiguous criteria for the identification of the photoreceptor. The most promising approach for an isolation would be a photoreceptor mutant and we described the properties those mutants should have. Until now there is no firm evidence that the photomutants, *madA* or *madB* are defective in the photoreceptor.

Despite some contradictory reports the method to trace the photoreceptor by means of light induced absorbance changes still seems promising, since photomutant specific alterations were found. This shows that the observed absorbance changes are likely to be related to the physiological event of light transduction. Even if this method should not lead to the identification of the photoreceptor, it may at least give some valuable information on the biochemical steps after the excitation of the photoreceptor. It is of vital importance to understand these early biochemical steps since one can expect that the adaptation mechanism resides in these early steps. Evidence for this comes from the fact that the photomutants, *madB* and *madC* are adaptation defective and *madC* at least can not be a photoreceptor mutant.

At the moment nothing is known about the molecular nature of the phenomenon of adaptation, i.e., range adjustment. The changes of adaptation levels to light intensity changes have been experimentally determined for the whole intensity range for the wild type and photomutants as well. The adaptation output seems identical in phototropism and light-growth response showing that most likely the same molecular mechanism underlies the two different light reactions. At the moment one is in need of a mathematical model which satisfactorily describes the various adaptation phenomena in wild type and mutants.

Later steps after the adaptation output involve transient changes of cAMP levels and a mutant with a defect in the *madD* gene turned out to have an altered phosphodiesterase. This *madD* mutant is the first behavioral mutant of *Phycomyces* whose biochemical nature is known.

The genetics which has been very helpful in a formal dissection of the light transduction pathway will have to be extended further with emphasis on the isolation of photoreceptor mutants. With respect to photodifferentiation it would be of great interest to isolate mutants which are defective in photodifferentiation while preserving their phototropical normality. These mutants should broaden our knowledge about the steps which phototropism and photodifferentiation do not have in common. In this respect the role of β-carotene in photophorogenesis and photocarotenogenesis deserves further investigation as both processes seem to be β-carotene dependent. In comparison with the light-growth response and phototropism the whole field of photodifferentiation of *Phycomyces* has been investigated to a rather small extent. The physiology of photocarotenogenesis and photophorogenesis has been sufficiently worked out to begin with biochemical studies and new efforts have to be made in this direction.

Acknowledgement. We are indebted to Drs. Thomas Trautner, Werner Schmidt, and John Reeve for many valuable discussions and suggestions. We thank Paul Manning and John Reeve for making the manuscript readable.

J. References

Abeles, F.B.: Ethylene in plant biology. New York: Academic Press 1973
Alpern, M.: J. Physiol. *217*, 447 (1971)
Bergman, K., et al.: *Phycomyces*. Bact. Rev. *33*, 99 (1969)
Bergman, K.: Planta *107*, 53 (1972)
Bergman, K., Eslava, A.P., Cerdá-Olmedo, E.: Mol. Gen. Genet. *123*, (1973)
Berns, P.S., Vaughn, J.R.: Biochem. Biophys. Res. Com. *39*, 1094 (1970)
Blalock, J.E., Gifford, G.E.: Proc. Natl. Acad. Sci. U.S.A. *74*, 5382 (1977)
Bonner, J.T., Dodd, M.R.: Devlop. Biology *5*, 344 (1962)
Bramley, P.M., Davies, B.H.: Phytochem. *15*, 1913 (1976)
Buder, J.: Ber. Dtsch. Bot. Ges. *36*, 104 (1918)
Castle, E.S.: J. Gen. Physiol. *14*, 701 (1931)
Castle, E.S.: J. Gen. Physiol. *17*, 751 (1934)
Castle, E.S.: 224 (1935)
Castle, E.S.: J. Cell. and Comp. Physiol. *8*, 493 (1936)
Castle, E.S.: Science *133*, 1424 (1961)
Castle, E.S.: J. Gen. Physiol. *49*, 925 (1966)
Cerdá-Olmedo, E.: Ann. Rev. Microbiol. *31*, 535 (1977)
Cohen, R., Delbrück, M.: J. Cell. Comp. Physiol. *52*, 361 (1958)
Cohen, R. J.: Life Sci. *15*, 289 (1974a)
Cohen, R. J.: Nature *251*, 144 (1974b)
Cohen, R. J., et al.: J. Gen. Physiol. *66*, 67 (1975)
Cohen, R J.: Plant Sci. Lett. *13*, 315 (1978)
Cohen, R. J., Atkinson, M.M.: Biochem. Biophys. Res. Com. *83*, 616 (1978)
Curry, G.M., Gruen, H.E.: Nature *179*, 1028 (1957)
Curry, G.B., Gruen, H.E.: Proc. Natl. Acad. Sci. U.S.A. *45*, 797 (1959)
Delbrück, M., Reichardt, W.: Cellular Mechanism in Differentiation and Growth. pp. 3–44. Princeton: Princeton University Press 1956
Delbrück, M., Shropshire, W.: Plant. Physiol. *35*, 194 (1960)
Delbrück, M., Varjú, D.: J. Gen. Physiol. *44*, 1177 (1961)
Delbrück, M., Katzir, A., Presti, D.: Proc. Natl. Acad. Sci. U.S.A. *73*, 1969 (1976)
De Luca L., et al.: Arch. Biochem. Biophys. *145*, 332 (1971)

Dennison, D.S.: Nature *184*, 2036 (1959)
Dennison, D.S.: J. Gen. Physiol. *45*, 23 (1961)
Dennison, D.S.: J. Gen. Physiol. *47*, 651 (1964)
Dennison, D.S.: J. Gen. Physiol. *48*, 393 (1965)
Dennison, D.S., Bozof, R.P.: J. Gen. Physiol. *62*, 157 (1973)
Dennison, D.S., Foster, K.W.: Biophys. J. *18*, 103 (1977)
Elfving, F.: Bot. Notiser *4*, 105 (1881)
Eslava, A.P., Alvarez, M.I., Cerdá-Olmedo, E.: Eur. J. Biochem. *48*, 617 (1974)
Eslava, A.P., Cerdá-Olmedo, E.: Plant Sci. Lett. *2*, 9 (1974)
Eslava, A.P., et al.: Molec. Gen. Genet. *147*, 235 (1976)
Fischer, E.P., Thomson, K.S.: J. Biol. Chem. (1979), in press
Foster, K.W., Lipson, E.D.: J. Gen. Physiology *62*, 590 (1973)
Galland, P., Russo, V.E.A.: Photochem. Photobiol. (1979a), in press
Galland, P., Russo, V.E.A.: Planta (1979b), in press
Garton, G.A., Goodwin, T.W., Liginsky, W.: Biochem. J. *48*, 154 (1951)
Goodwin, T.W., Griffith, L.A., Modi, V.V.: Biochem. J. *62*, 259 (1956)
Hsu, W.J., Ailion, D.C., Delbrück, M.: Phytochem. *13*, 1463 (1974)
Jaffe, L.F.: J. Gen. Physiol. *43*, 897 (1960)
Jan, Y.N.: J. Biol. Chem. *249*, 1973 (1974)
Jesaitis, A.J.: J. Gen. Physiol. *63*, 1 (1974)
Johnson, D.L., Gamow, R.I.: J. Gen. Physiol. *57*, 41 (1971)
Kaufmann, D.G., et al.: Science *177*, 1105 (1972)
Kurtin, W.E., Song, P.S.: Photochem. Photobiol. *7*, 263 (1968)
Lafay, J.F., Matricon, J., Bodéré, C.: Physiol. Rég. *13*, 259 (1975)
Lipson, E.D.: Biophys. J. *15*, 989 (1975a)
Lipson, E.D.: Biophys. J. *15*, 1013 (1975b)
Lipson, E.D.: Photomutants. Biophys. J. *15*, 1033 (1975c)
Lipson, E.D., Presti, D.: Photochem. Photobiol. *25*, 203 (1977)
Mackinney, G., et al.: J. Am. Chem. Soc. *74*, 3436 (1952)
Meistrich, M.L., Fork, R.L., Matricon, J.: Science *169*, 370 (1970)
Meissner, G., Delbrück, M.: Plant Physiol. *43*, 1279 (1968)
Miki, N., et al.: Proc. Nat. Acad. Sci. U.S.A. *70*, 3820 (1973)
Moore, T.: In: The vitamins. 2nd ed. Sebrell, W.H., Harris, R.S. (eds.). pp. 245–266. New York: Academic Press 1967
Muñoz, V., Butler, W.L.: Plant Physiol. *55*, 421 (1975)
Ootaki, T., Crafts-Lighty, A., Delbrück, M.: Mol. Gen. Gent. *121*, 57 (1973)
Ootaki, T., Fischer, E.P., Lockhard, P.: Molec. Gen. Genet. *131*, 233 (1974)
Pilet, P.E.: Experientia XII/4, 148 (1956)
Poff, K.L., Butler, W.L., Loomis, W.L., Jr.: Proc. Natl. Acad. Sci. U.S.A. *70*, 813 (1973)
Poff, K.L., Butler, W.L.: Nature *248*, 799 (1974)
Presti, D., Hsu, W.-J., Delbrück, M.: Photochem. Photobiol. *26*, 403 (1977)
Rau, W.: Ber. Dtsch. Bot. Ges. *88*, 45 (1975)
Reau, P.: Planta *108*, 153 (1972)
Reichardt, W., Varjú, D.: Z. Für Phys. Chem. Neue Folge *15*, 297 (1958)
Reyes, P., Chichester, C.O., Nakayama, T.O.M.: Biochim. Biophys. Acta *90*, 578 (1964)
Roelofsen, P.S.: Rec. Trav. Bot. Neer. *42*, 73 (1950)
Rushton, W.A.H., Henry, G.H.: Vis. Res. *8*, 617 (1968)
Russo, V.E.A.: J. Bact. *130*, 548 (1977a)
Russo, V.E.A.: Plant. Sci. Lett. *10*, 373 (1977b)
Russo, V.E.A., Halloran, B., Gallori, E.: Planta *134*, 61 (1977)
Sandmann, G., Hilgenberg, W.: Biochem. Physiol. Pflanzen *172*, 401 (1978)
Sargent, M.L., Briggs, W.R.: Plant. Physiol. *42*, 1504 (1967)
Schrott, E.L., Rau, W.: Planta *136*, 45 (1977)
Shropshire, W., Jr.: Science *130*, 336 (1959)
Shropshire, W., Jr.: J. Gen. Physiol. *45*, 949 (1962)

Shropshire, W., Jr.: First European Biophysics Congress. Broda, E., Locker, A., Springer-Lederer, H., (eds.) Verlag der Wiener Med. Akad., pp. 111–114 (1971)
Siødmiak, J., Frackowiak, D.: Photochem. Photobiol. *16*, 173 (1972)
Song, P.L., Moore, T.A.: Photochem. Photobiol. *19*, 435 (1974)
Thornton, R.M.: Plant Physiol. *49*, 194 (1972)
Thornton, R.M.: Plant. Physiol. *51*, 570 (1973)
Thornton, R.M.: Amer. J. Bot. *62*, 370 (1975)
Tu, J.C., Malhotra, S.K.: J. Histochem. Cytochem. *21*, 1041 (1973)
Uebelmesser, E.R.: Arch. Mikrobiol. *20*, 1 (1954)
Ulane, R.E., Cabib, E.: J. Biol. Chem. *251*, 3367 (1976)
Varjú, D., Edgar, L., Delbrück, M.: J. Gen. Physiol. *45*, 47 (1961)

Light Perception and Sensory Transduction in Photosynthetic Prokaryotes

Wilhelm Nultsch and Donat-Peter Häder

Fachbereich Biologie der Universität Marburg, Lahnberge, D-3550 Marburg/Lahn, Federal Republic of Germany

Table of Contents

This article elucidates molecular mechanisms in light perception, stimulus transformation and signal transmission in photosynthetic prokaryotes. Special emphasis is put on the distinction between various coupling between different light responses and photosynthesis.

I. Introduction

In photosynthetic organisms the bulk of the pigments absorbing near u.v., visible and infrared radiation, belong to the photosynthetic apparatus and are either integrated into the thylakoid membranes (chlorophylls, carotenoids) or closely associated with them, as in the case of phycobiliproteins (phycobilisomes). However, besides photosynthesis, developmental processes and motile responses can also be influenced by light. In some cases they are directly coupled with photosynthesis, while in others the photosynthetic pigments are used for light absorption, although the excitation energy seems not to be utilized via photosynthesis. Finally, special light-absorbing systems completely independent of the photosynthetic apparatus are developed.

A great variety of light-triggered processes exists even in phototrophic prokaryotes (photosynthetic bacteria, Cyanobacteria = Cyanophyceae = blue-green algae). In the case of movement we distinguish between

1. photokinesis, i.e., the effect of light intensity on the speed of movement,
2. phototaxis, i.e., the orientation of movement towards light, and
3. photophobic response, i.e., a transient stop or reversal of movement caused by a sudden temporal change in light intensity [100,102].

Photocontrolled developmental processes are

1. photomorphogenesis, i.e., the effect of light on growth and formation of the organisms [67],
2. development of the photosynthetic apparatus, and
3. chromatic adaptation, i.e., the change in the ratio of phycobiliproteins to one another and/or chlorophyll a, depending on the spectral composition of the incident radiation in blue-green algae [7].

Regardless of whether developmental or motile processes are concerned, the reaction chain can be subdivided into the following steps:

1. stimulus perception, i.e., the absorption of radiation by the photoreceptor pigments,
2. stimulus transformation, i.e., the conversion of the absorbed radiant energy into a transmittable form,
3. signal transmission to the effector, which in most organisms is located some distance from the photoreceptor system,
4. the developmental or motile response, which in turn might consist of several subsequent steps.

The molecular basis of most of these processes is not yet clear. Most of what we know concerns the absorption of quanta by the photoreceptor pigments, while stimulus transformation and signal transmission have as yet been investigated only in part and in a very few organisms. In blue-green algae, not even the movement mechanism is clear, so that any model of the motor response is highly speculative.

II. Photoreceptors

To be effective, light must be absorbed by molecules with characteristic bands, the photoreceptor pigments. Photosynthetic pigments are involved in the absorption of the active light in all photomotile responses of photosynthetic prokaryotes investigated thus far. There are some action spectra indicating the participation of all photosynthetic pigments (chlorophylls, carotenoids and phycobiliproteins), while in other processes one or the other group predominates. In photocontrolled developmental processes, the so-called phycochromes seem to play an important role (see below). Flavin-mediated photoresponses will be covered in the chapter by Doughty and Diehn.

1. Bacteriochlorophylls

Bacteriochlorophylls a, b, c, d, and e have been isolated from photosynthetic bacteria (see review[113]). Since photomotile responses of Chlorobacteriaceae have not yet been investigated, we are limited to the studies of Rhodospirillaceae and Chromatiaceae, which contain only one bacteriochlorophyll, either a or b[114]. The physical and chemical properties of these pigments have recently been described by Gingras[34], Thornber et al.[132], Olson[111], and Mauzerall[76]. As shown in the formula, bacteriochlorophyll a is a 2-desvinyl-2-acetyl-3,4-dihydrochlorophyll.

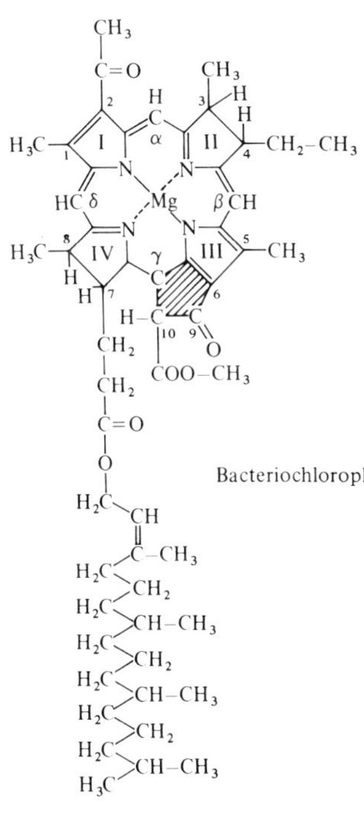

Bacteriochlorophyll

a) Photophobic Response

The role of bacteriochlorophyll in the absorption of photophobically active light was already demonstrated by Engelmann [27] in the so-called *Bacterium photometricum*, probably a *Chromatium* species. This was confirmed later by other investigators who measured action spectra of the photophobic responses in several different species (see reviews by Clayton [17] and Nultsch [95,97]). One of the spectra measured by Clayton [15] is shown in Fig. 1. This photophobic action spectrum agrees well with the in vivo absorption spectrum in *Rhodospirillum rubrum*, indicating that light absorbed by bacteriochlorophyll a is active. Some smaller peaks or shoulders in the shorter wavelengths region (400–600 nm) indicate the participation of carotenoids in the active light absorption (see below).

Because of the striking similarity of the action spectra of photosynthesis and the photophobic response, Manten [73] concluded that photophobic responses are caused by sudden changes in the rate of photosynthesis. But Clayton's experiments [16] gave evidence that even if photosynthesis is saturated, a sudden change in light intensity can cause a transient change in the steady state of photosynthesis, thereby initiating a photophobic response. As a general hypothesis, Links [71] has suggested that phobic responses are caused by sudden changes in the ATP supply from photophosphorylation to the motor apparatus. However, this is obviously not true in the case of the photophobic response of purple bacteria. Throm [133], studying the effects of inhib-

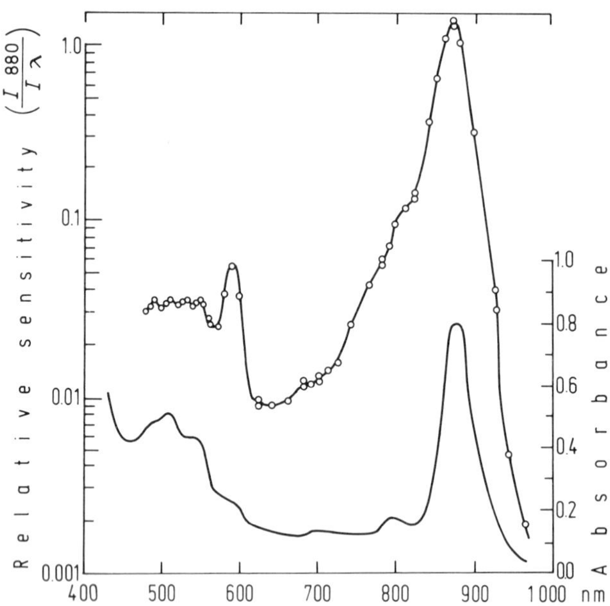

Fig. 1. Photophobic action spectrum *(open circles, solid line)* and in vivo absorption spectrum *(solid line)* of *Rhodospirillum rubrum*. *Abscissa:* wavelength in nm; *Ordinates:* relative sensitivity and absorbance respectively (after Clayton [15], from Haupt [57]), modified

itors on the photophobic response of *Rhodospirillum rubrum*, has shown that blockers of the photosynthetic electron transport (with the exception of DCMU) inhibit photophobic responses, whereas uncouplers such as dinitrophenol (DNP) have no effect (Fig. 2). A selection of inhibitors and uncouplers is listed at the end of this chapter. Clayton[17)] proposed the hypothesis that photophobic responses in purple bacteria are "mediated through the development of an excitatory state which is transmitted to the locomotor areas, causing a coordinated motor response".

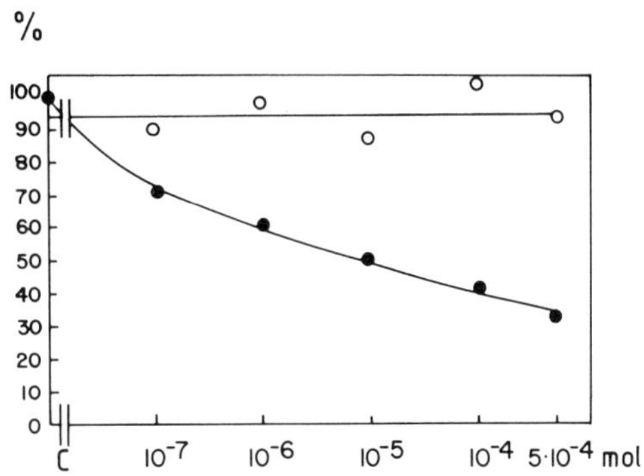

Fig. 2. Effect of DNP on photokinesis *(closed circles)* and photophobic response *(open circles)* of *Rhodospirillum rubrum*. *Abscissa:* DNP concentration in mol; *Ordinate:* response in % of the uninhibited control (after[133)])

b) Photokinesis

Photokinesis in purple bacteria was observed first by Engelmann[26–28)]. These organisms became motionless in the dark but resumed movement after the onset of irradiation. Their speed increased with greater light intensity. Engelmann concluded that a pool of a certain substance may be produced by irradiation, which is necessary for movement and is consumed in the dark. Throm[133)], in measuring the photokinetic action spectrum of *Rhodospirillum rubrum*, found that it also resembled the absorption spectrum of the cells and the action spectrum of photosynthesis. This means that the photokinetically active radiation is absorbed by the photosynthetic pigments, mainly bacteriochlorophyll a, although a relatively high peak at 510 nm and a shoulder at 550 nm indicate a stronger participation of spirilloxanthin in photokinesis than in photosynthesis. Since photokinesis of *Rhodospirillum* is sensitive to uncouplers and inhibitors of the photosynthetic electron transport as well, it can be concluded that photokinesis is coupled with photophosphorylation, as in blue-green algae (see below).

c) Development

In purple bacteria investigated thus far, the development of the photosynthetic apparatus which is controlled by the oxygen partial pressure is not influenced by light. However, the prevalent light intensity modifies the morphogenetic process of formation of intracytoplasmic membranes. Light intensity also controls the total bacteriochlorophyll concentration per cell and per membrane protein, as well as regulating the ratio of the steady state concentrations of the various photosynthetic pigments, i.e., the ratio of reaction center bacteriochlorophyll: light-harvesting bacteriochlorophyll(s) and carotenoids [30,124]. Since the action spectrum of bacteriochlorophyll synthesis is essentially identical to the absorption spectrum of photosynthetic pigments [24], light acts via the photosynthetic apparatus itself. Further evidence has been presented by Göbel [37], who has shown that all effects of light intensity changes on the biosynthetic apparatus are mediated by bacteriochlorophyll. Thus, no special photomorphogenetic pigment system seems to be developed in photosynthetic bacteria. It has been suggested that the effect of light is due to the production of a corepressor interacting with the reaction center bacteriochlorophyll. For more information see the review by Drews [23].

2. *Chlorophyll a*

Unlike the photosynthetic apparatus of photosynthetic bacteria, that of cyanobacteria consits of two photosystems, PS I and II, connected by an electron transport chain. The only chlorophyll present is chlorophyll a, and, therefore, chlorophylls b–d are not of interest in this article. Chlorophyll a is the principal constituent of PS I. Twenty per cent of isolated pigment-protein complexes contain one P700 per 20–30 chlorophyll a molecules; the other 80% contain only chlorophyll a [20]. The physical and chemical properties of chlorophyll a and its role in photosynthesis have recently been described by Meeks [77], Mauzerall [75], Hoch [60], Butler [10], and other authors of the Encyclopedia of Plant Physiology NS Vol. 5.

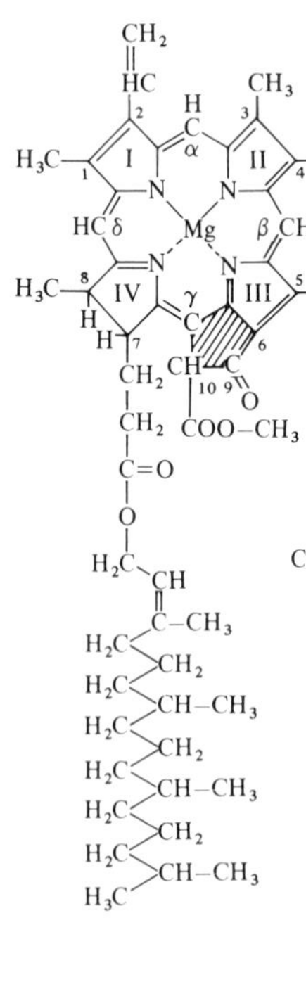

Chlorophyll a

a) Photokinesis

Photokinesis was the first photomotile response of blue-green algae in which the participation of chlorophyll a in the active light absorption was demonstrated[87,89,103]. The photokinetic action spectra measured in some *Phormidium* species either resemble a "pure" chlorophyll a spectrum (Fig. 3) or indicate a slightly increased activity of phycobiliproteins. Therefore, it has been concluded that in the genus *Phormidium*, chlorophyll a is the main photoreceptor in the visible region, and that the significant effectiveness of the far red and the near infrared is also due to PS I activity. This suggests that in these forms the acceleration of movement by light is essentially the result of an energy supply from cyclic photophosphorylation driven by PS I.

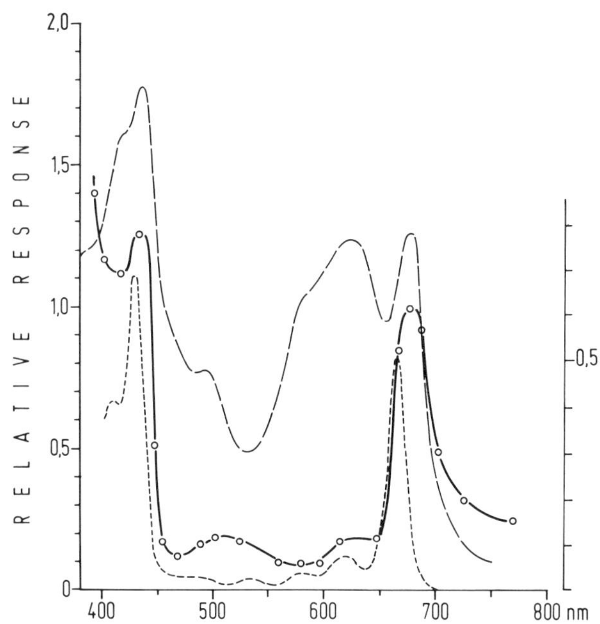

Fig. 3. Photokinetic action spectrum (*circles* and *solid line*) and in vivo absorption spectrum (*long dashed line*) of *Phormidium ambiguum*. Absorption spectrum of chlorophyll a (*short dashed line*). *Abscissa:* wavelength in nm; *Ordinates:* relative response and absorbance respectively (modified after[103])

The photokinetic action spectra of other blue-green algae, such as *Anabaena variabilis*, *Pseudanabaena catenata, Oscillatoria tenuis, Nostoc muscorum* and *Spirulina subsalsa*, are consistent with those of *Phormidium* in that radiation around 435 and 675 nm, indicating chlorophyll a absorption, is active. However, some of them display a stronger activity in the range between 450 and 500 nm, absorbed mainly by carotenoids; in others, though, strong or maximal activity is found between 615 and

630 nm, indicating C-phycocyanin as photoreceptor pigment. In his interpretation of these results, Nultsch[96,97] has suggested that photokinesis is due to an additional energy supply of photosynthetic phosphorylation to the locomotor apparatus, and that cyclic phosphorylation and/or a phosphorylation coupled with the non-cyclic electron transport can serve as energy sources. These conclusions were supported by the effects of photosynthetic inhibitors, uncouplers, and redox systems on photokinesis of blue-green algae[91–94,103,104]. Thus, photokinesis is a typical photocoupling process.

b) Photophobic Response

A comparison of the action spectrum of the photophobic response measured by Nultsch[88] in *Phormidium uncinatum* with the low termperature absorption spectrum[39] shows maximal activity in the range of around 560 nm absorbed by C-phycoerythrin (Fig. 4), but also a distinct peak of red chlorophyll a with a maximum of around 675 nm. However, the effectiveness of blue light is far out of proportion to the absorption by the Soret band of chlorophyll a. The same is true in the action spectrum of *Phormidium ambiguum*. This species contains C-phycocyanin but no C-phycoerythrin, and, consequently, the maximum occurs at around 615 nm. Since these action spectra essentially agree with the photosynthetic spectra measured by Duysens[25], Haxo and Norris[58], and Nultsch and Richter[105] in some *Oscillatoria* and *Phormidium* species, it has been concluded that in blue-green algae, photophobic responses are also mediated by the photosynthetic apparatus, as in purple bacteria.

Investigating the effects of photosynthesis inhibitors, uncouplers, and redox systems, Nultsch[90–94,96,97], Nultsch and Jeeji-Bai[104], and Häder[41] found that inhibi-

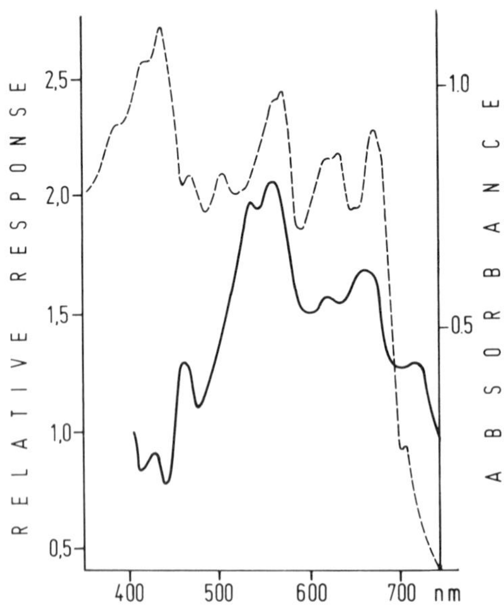

Fig. 4. Photophobic action spectrum *(solid line)* and low temperature absorption spectrum *(dashed line)* of *Phormidium uncinatum*. *Abscissa:* wavelength in nm; *Ordinates:* relative response and absorbance respectively (after Nultsch[88], Häder[39], modified)

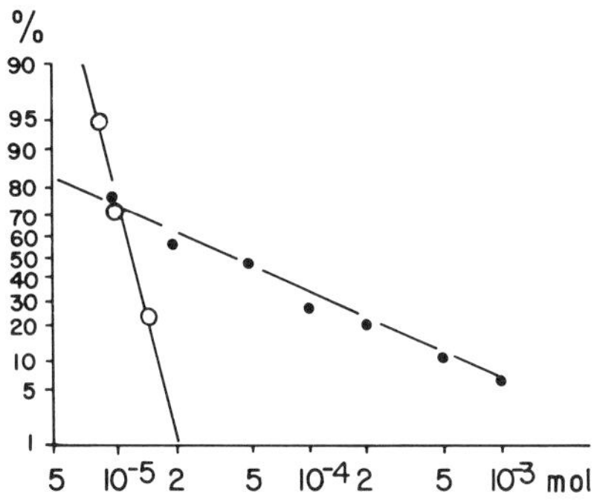

Fig. 5. Effect of o-phenanthroline on photophobic response *(open circles)* and photokinesis *(closed circles)* of *Phormidium uncinatum*. *Abscissa:* o-phenanthroline concentration in mol; *Ordinate:* response in % of the uninhibited control (after Nultsch and Jeeji-Bai [104])

tors of the photosynthetic electron transport decrease the photophobic reactivity. As shown in Fig. 5, the photophobic response of *Phormidium uncinatum* is very sensitive to o-phenanthroline, whereas photokinesis is only slightly impaired. The uncoupler experiments yielded equivocal results, since most of these substances also inhibit oxidative phosphorylation and, hence, movement in the dark, thereby preventing the organisms from entering the light trap. Redox systems inhibit photophobic responses the closer their midpoint potential approaches that of the natural electron acceptors of both photosystems. From these results, Nultsch [93] concluded that in this organism, photophobic responses are caused by sudden changes in the steady state of the photosynthetic electron transport, provided that these changes exceed a certain threshold value. This is contrary to the hypothesis of Links [71], according to which photophobic responses are initiated by a decrease in the energy supply to the motor apparatus (see above).

c) Phototaxis

While phototactic action spectra measured in some *Phormidium* species indicate that chlorophyll a is not involved in the absorption of phototactically active light (see below), the phototactic action spectrum of *Anabaena variabilis* [106] shows slight activity around 440 nm and a distinct peak at around 670 nm, both indicating chlorophyll a. Since blockers of the photosynthetic electron transport, such as DCMU and DBMIB, (see below) do not affect phototactic orientation, the active light seems not to be utilized via the photosynthetic electron transport chain (for further information see below).

3. Carotenoids

Purple bacteria and blue-green algae both contain a number of carotenoid pigments. The carotenoids of the purple bacteria belong to the following series (see review by Pfennig [113]): normal spirilloxanthin series, rhodopinal branch of spirilloxanthin series, alternative spirilloxanthin series, okenon series, and isorenieratene series. From blue-green algae a great variety of carotenoids has been isolated [31,38,84]. β-Carotene seems to be universal, and myxoxanthophyll and oscilloxanthin appear to be specific to this group. As mentioned above, action spectra studies with the purple bacterium *Rhodospirillum rubrum* have shown that carotenoids are active as photoreceptors in photokinesis, as well as in the photophobic response. In blue-green algae the principal photosynthetically active carotenoids are β-carotene and zeaxanthin [130]. Since it has been shown that photokinesis of blue-green algae is coupled with photophosphorylation, and the photophobic response with the photosynthetic electron transport (see above), the participation of carotenoids in the absorption of photokinetically and photophobically active light can be expected. This is substantiated by most of the action spectra measured so far.

Moreover, carotenoids also participate in the absorption of the photoactically active light in some *Phormidium* species (see below). Since in this case the participation of chlorophyll a is not evident in the action spectrum, the absorbed energy does not seem to be utilized via the photosynthetic apparatus. It is possible, though, that the bulk of carotenoids is associated with a few specialized chlorophyll molecules serving as energy traps. Their response might not be detectable in the action spectrum due to their low concentration, as in the case of the peridinin-chlorophyll a-complex in some dinoflagellates [61,127,128]. However, no concepts of the transduction of the information to the locomotor apparatus have as yet been proposed.

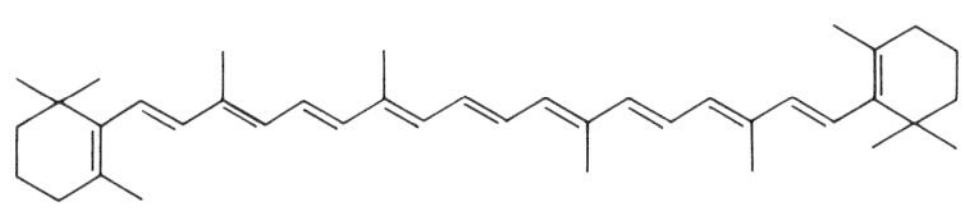

β–Carotene

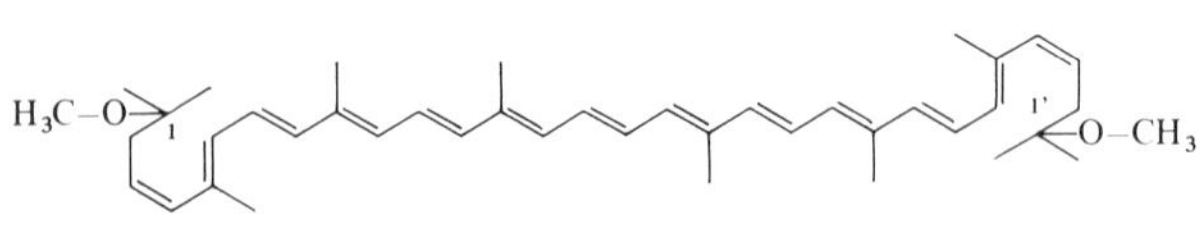

Spirilloxanthin

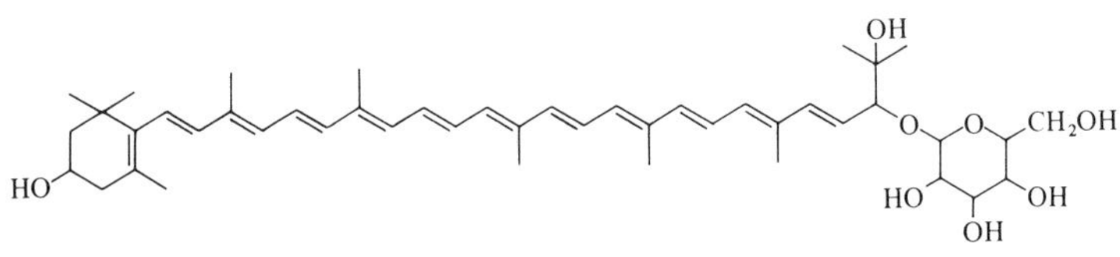

Myxoxanthophyll

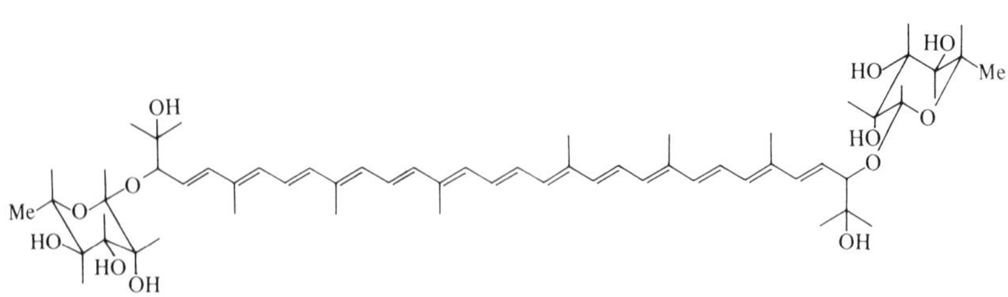

Oscillaxanthin

4. *Phycobiliproteins*

C-phycoerythrin, C-phycocyanin and allo-phycocyanin are the major accessory pigments of the blue-green algae. Located in the phycobilisomes, they collect energy which is transferred to the chlorophyll a molecule of the reaction center of PS II through the sequence C-phycoerythrin → C-phycocyanin → allo-phycocyanin (see review by Krogmann[63]). The chemical and physical properties of the phycobiliproteins have been reviewed recently by Chapmann[13], Rüdiger[118,119], and O'Carra and O'hEocha[107], and the structure of the light-harvesting pigment-protein complexes by Zuber[139]. The chromophore groups, phycoerythrobilin and phycocyanobilin respectively, are shown in the formulas below.

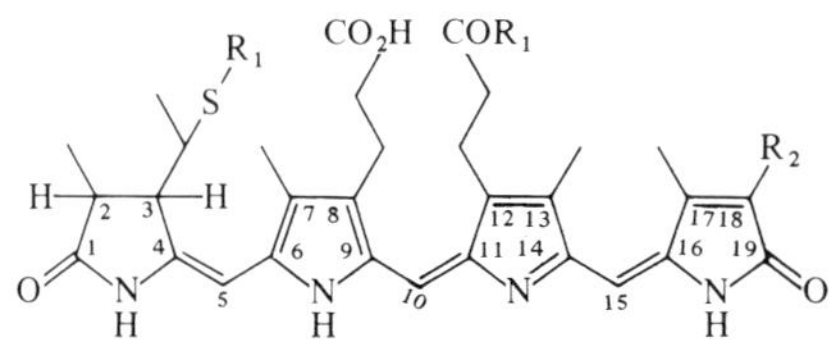

Phycocyanobilin R_1 = Protein; R_2 = C_2H_5

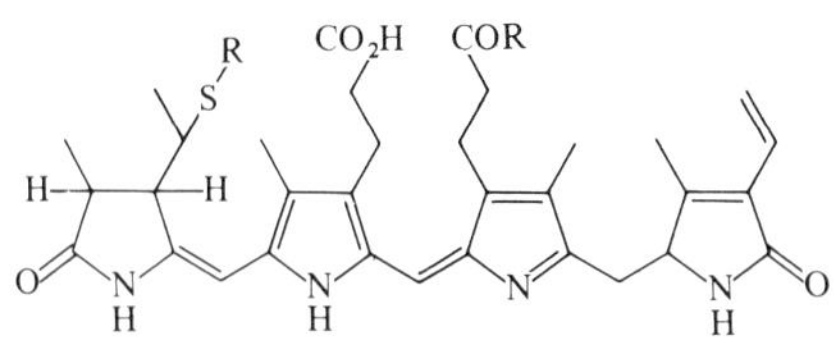

Phycoerythrobilin R = Protein

a) Photokinesis

The participation of the phycobiliproteins in the absorption of photokinetically active light has been demonstrated above. Peaks of around 565 and 615 nm in the action spectra indicate the involvement of C-phycoerythrin and C-phycocanin. These pigments transfer energy to the reaction center of PS II and suggest the participation of the non-cyclic electron transport and coupled phosphorylation.

b) Photophobic Response

As mentioned above, in all photophobic action spectra measured so far additional peaks and/or shoulders occur at 565 and/or 615 nm, besides the chlorophyll a peaks, indicating the participation of phycobiliproteins and, hence, PS II in the active light absorption (Fig. 4). Moreover, activity of both PS I and PS II was confirmed by inhibitor experiments (see reviews by Nultsch[100], Nultsch and Häder[102], and Häder[47]) and also with the aid of a dual wavelength technique[39,40,101], described in detail below.

These experiments show that, as in the case of chlorophyll a and the carotenoids, the energy absorbed by the phycobiliproteins is utilized via the photosynthetic apparatus; furthermore, they provide evidence that photophobic responses in blue-green algae are caused by sudden changes in the steady state of the photosynthetic electron transport, especially the non-cyclic one.

c) Phototaxis

Phototactic action spectra of *Phormidium autumnale* and *Phormidium uncinatum*, measured by Nultsch[86,89], show prominent maxima in the absorption range of C-phycoerythrin and smaller, but distinct, peaks in the absorption range of C-phycocyanin. Red light absorbed by chlorophyll a is not active, while in the blue range absorbed by the Soret band, the action spectrum shows a minimum (Fig. 6). Nultsch[87] concluded that biliproteins are photoreceptors of phototaxis, but independently of the photosynthetic electron transport and phosphorylation.

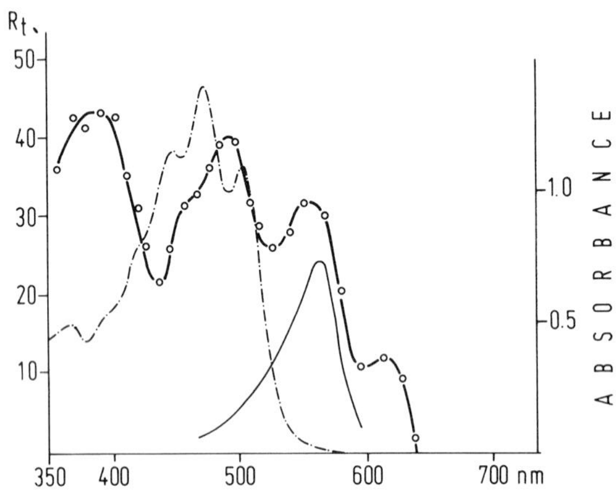

Fig. 6. Phototactic action spectrum (*circles* and *heavy solid line*) of *Phormidium uncinatum*. For comparison the absorption spectra of C-phycoerythrin *(fine solid line)* and myxoxanthophyll *(dashed-dotted line)* are drawn. *Abscissa:* wavelength in nm; *Ordinates:* phototactic reaction value R_t in relative units and absorbance respectively (modified after Nultsch[89])

The action spectrum of positive and negative phototaxis of *Anabaena variabilis* was measured recently[106]. This species contains no C-phycoerythrin. Accordingly, maximum activity is found at around 615 nm (Fig. 7). In addition, in this form a second maximum occurs at around 675 nm, and a third small, but distinct, one at 440 nm, both indicating that chlorophyll a is also involved in the active light absorption (see above). The utilization via photosynthesis, however, could be excluded in this case, since the trichomes oriented themselves perfectly well to the light direction in the presence of photosynthetic inhibitors, such as DCMU and DBMIB, at concentrations in which the photosynthetic oxygen evolution was almost completely inhibited.

The situation is further complicated by the finding that in negative phototaxis, radiation between 500 and 560 nm and above 700 nm is effective in addition to the wavelengths, which are active in the positive response. Nultsch et al.[106] have discus-

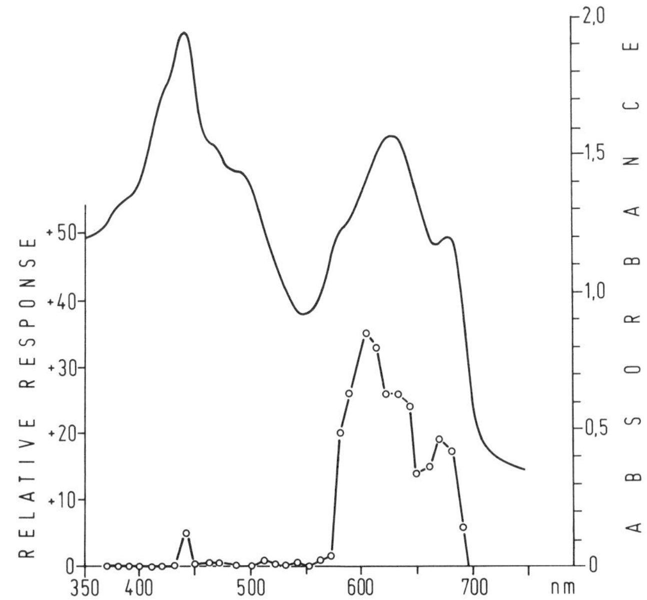

Fig. 7. Action spectrum of positive phototaxis (*circles* and *solid line*) and in vivo absorption spectrum *(solid line)* of *Anabaena variabilis. Abscissa:* wavelength in nm; *Ordinates:* phototactic effect in relative units and absorbance respectively (after Nultsch et al.[102])

sed the possibility of whether the photosynthetic pigments act as photoreceptors or only as a screen producing an absorption gradient between the illuminated and shaded sides. This question, however, cannot yet be decided.

5. *Phycochromes*

Photomorphogenetic effects in blue-green algae were first reported by Lazaroff and Schiff[68], who observed that in *Nostoc muscorum*, filamentous growth was induced by light between 600 and 700 nm, and that this induction was reversed by light between 500 and 600 nm. At the same time, Fujita and Hattori[33] measured the action spectrum of the conversion of C-phycocyanin into C-phycoerythrin precursors in *Tolypothrix tenuis* and found activity between 480 and 580 nm, with a maximum at 541 nm. Later, Lazaroff[66] demonstrated that no correlation exists between photocontrol of development and photosynthetic processes.

Since then several other effects of light on morphogenesis, pigment synthesis, and chromatic adaptation have been observed which have been reviewed by Lazaroff[67] and Bogorad[7]. The existence of separate photoreceptive systems for chromatic adaptation (adaptochromes) and photomorphogenesis (morphochromes) has been suggested by Bogorad and coworkers (c.f., Bogorad[7]). Some of these pigments are suspected to be hemoproteins and others biliproteins.

Recently, photoreversible pigments, called phycochromes, were extracted from blue-green algae by Björn and Björn[5,6] and Björn[3,4]. Four types were found:

Phycochrome a, extracted from *Tolypothrix distorta, Phormidium luridum, Nostoc muscorum* and *Anacystis nidulans*, has one form absorbing maximally at about 590 nm, formed in red light, and another one absorbing at 570 nm, formed in yellow-green and blue-green light.

Phycochrome b, also found in *Tolypothrix*, occurs in two forms absorbing maximally at 510 and 570 nm respectively, and produced in yellow-green and blue-green light respectively.

Phycochrome c, occurring in *Nostoc* and probably also in *Tolypothrix*, has one form absorbing at around 650 nm and another, absorbing weakly in the green, produced in green and red light respectively.

Phycochrome d, extracted also from *Tolypothrix*, shows an absorbance increase at 650 nm after irradiation with 650 nm, as well as an absorbance decrease at the same wavelength after irradiation with 610 nm light.

As Björn and Björn[5] have shown, the difference spectrum of phycochrome c and the action spectrum for induction of filamentous growth of *Nostoc muscorum*[68] are similar insofar as the maxima are found at the same wavelength (Fig. 8). They suggest that phycochrome c might be the physiologically active pigment of this photomorphogenetic process. The photoreceptor for chromatic adaption in *Tolypothrix tenuis*[19], however, appears to have a significantly different spectrum (Fig. 8). The chemical nature of the phycochromes is not yet known. However, it has been shown by Ohki and Fujita[108,109] in *Tolypothrix tenuis* that phycocyanin and allophycocyanin obtain the ability for photoresponsiveness when their protein conformation, probably around the chromophore site, is modified.

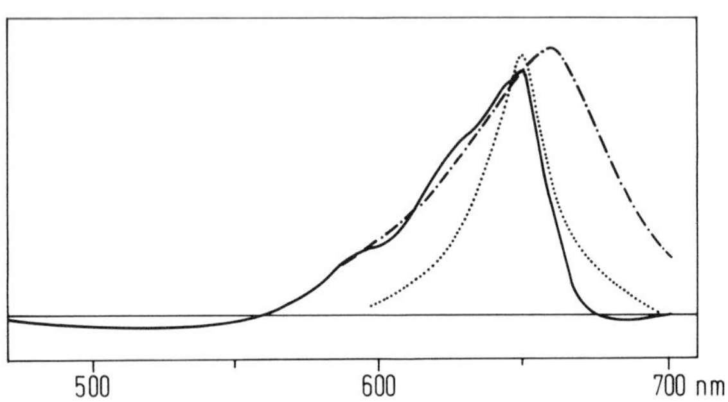

Fig. 8. Difference spectrum (green irradiated minus red irradiated) of phycochrome c *(solid line)*, action spectrum for induction of filamentous growth of *Nostoc muscorum (dotted line)*, and action spectrum for inhibition of C-phycoerythrin synthesis in *Tolypothrix tenuis (dashed-dotted line)*. *Abscissa:* wavelength in nm; *Ordinates:* relative absorbance change und relative action (after Björn and Björn[5])

III. Stimulus Transformation

1. Functions and Possible Mechanisms

Once a quantum has been absorbed by a photoreceptor pigment and utilized to trigger a primary photoevent (e.g., a conformational change of a molecule or a transition into an excited state), the energy must be converted in order to be transmitted to the final acceptor – in the case of photomovement, to the motor apparatus of the organism. This sequence of events has been called sensory transduction (cf.[43]). The energy of the absorbed quantum can be directly used to drive the following steps in the reaction chain, as in the case of photosynthesis; or, it can be amplified, for example, by gating a transmembrane channel in order to provide a directed ion or proton flux through the membrane (cf. review[56]).

2. Formation of Chemical Transmitters

As described above, photokinesis in photosynthetic bacteria and in blue-green algae is the result of an additional energy supply from photosynthetic phosphorylation. In blue-green algae, cyclic, non-cyclic, or pseudocyclic phosphorylation can contribute to the effect, though to a different extent in different species (Nultsch and Hellmann[103]). In *Rhodospirillum rubrum*, light can be substituted by externally applied ATP. 10^{-2} M ATP increases the speed of movement by the factor 3, compared to dark movement, though only after a considerable lag phase, and may be due to the uptake through the cytoplasmic membrane[133]. According to Larsen et al.[65], in bacteria the energy source for swimming is a high-energy intermediate rather than ATP. However, chemical energy stored in the form of ATP can be converted into this intermediate (cf. review by Sistrom[126]). Thus, whatever the final form of energy is here, the absorbed light energy is transformed into a chemical transmitter.

3. Electron Pool Hypothesis

As mentioned above, the photophobic response in blue-green algae is linked to photosynthesis as is photokinesis. Inhibitors of the photosynthetic electron transport impair the photophobic response, while uncouplers do not[98,99]. This indicates that the photophobic response is triggered by sudden changes in the rate of the photosynthetic electron flow. In order to elucidate the mode of coupling, a dual wavelength light trap method was used[40,49]. The homogeneously suspended organisms were evenly illuminated by a monochromatic background light from above, while from below, a small light field of the same or different wavelength was projected into the preparation. Depending on the wavelengths, this resulted in accumulations in, or dispersal of, organisms from the light trap.

When the trap light is utilized predominantly by photosystem II (e.g., by irradiation of 563 nm, which is absorbed by the main accessory pigment C-phycoerythrin), the organisms respond phobically upon leaving the light field under all background wavelengths, resulting in accumulations in the trap. Varying the background wavelength produced an action spectrum which may represent the photosynthetic action spectrum of photosystem I [101].

In a constant trap wavelength activating photosystem I (e.g., 700 nm, which is absorbed by P 700), the response depends on the background wavelength. In a background wavelength ranging from 450 to 630 nm, the organisms respond phobically upon leaving the light trap, thereby resulting in photoaccumulations; in background wavelengths above 630 nm and below 450 nm, however, they respond phobically upon entering the light field [40] and disperse. Varying the background wavelength produced an action spectrum similar to that of photosynthesis involving only photosystem II. When both photosystems are activated, the type of response depends on the ratio of quanta absorbed by photosystem I or II respectively. Similar results have been obtained by using a technique in which two monochromatic light fields of different wavelengths were irradiated next to each other [42].

These results were interpreted using the electron pool hypothesis: There is an electron pool situated in the linear photosynthetic electron transport chain between photosystems II and I (Fig. 9). A phobic response is triggered by a decrease in the flow rate through the pool. This can be accomplished in two ways:

1) When an organism leaves a photosystem II light trap, the flux into the electron pool decreases.

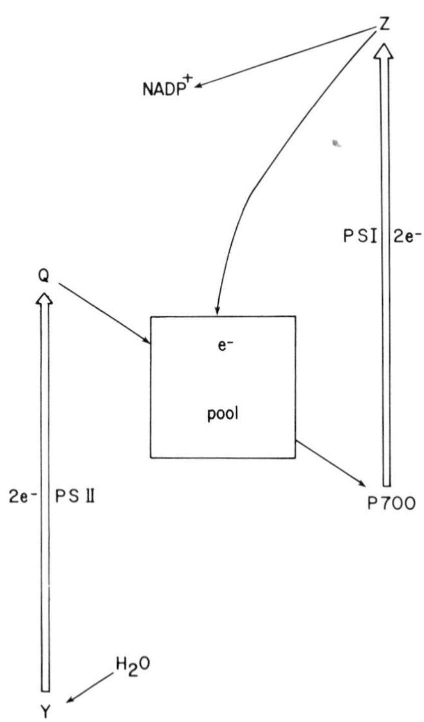

Fig. 9. Schematic diagram explaining the electron pool hypothesis (after Häder [40])

2) When a trichome moves into a photosystem I light trap, electrons are drained out of the pool. In both cases a phobic response is caused, but different patterns result, one being an accumulation in, the other a dispersal from, the light trap [101].

The molecular characterization of the electron pool has been revealed by the application of various electron transport inhibitors. DCMU impairs the non-cyclic photosynthetic electron transport in the immediate neighborhood of photosystem II (cf. Codd and Cossar [18]). In addition, DCMU inhibits the phobic response mediated by photosystem I light only in fairly high concentrations, while it blocks a photosystem II-mediated response quite specifically [41]. DBMIB, on the other hand, interrupts the non-cyclic photosynthetic electron transport chain immediately after plastoquinone by preventing the reoxidation of this redox system [85,125,134]. The effect of this inhibitor on the photophobic response is contrary to that of DCMU. The same is true for KCN and DSPD (disalicyliden propandiamin), which are believed to inhibit plastocyanin and ferredoxin respectively. These results indicate that plastoquinone is the site of the electron pool.

IV. Signal Transmission

1. Mechanisms for Signal Transmission

There are several possible mechanisms for providing an information transport between the photoreceptor and motor apparatus:

1) The primary photoevent may cause the production of a chemical transmitter, which either diffuses through the cytoplasm or is transported in a specific compartment of the cell. In photokinesis in blue-green algae, this transport takes several minutes, thereby indicating passive diffusion, since the organisms adapt their speed to a new level of illumination only after about 10 min [47].

2) The utilization of the piezoelectric effect has been discussed both for photosynthesis [11] and phototaxis [8]. This involves an organelle with quasi-cristalline properties as a transducer similar to the paraflagellar rod described for some euglenoid flagellates [115].

3) In a number of systems the generation of electrical potential changes has been suggested [21,54,55] and will be discussed in greater detail with regard to photophobic responses in blue-green algae (Sect. II, 4). Instead of one mechanism for signal transmission, different microorganisms seem to utilize different ones.

2. Transmission by Electrical Potential Changes

Calculations by Lee and Fitzsimons [69] have shown that even in bacteria diffusion of a small molecule is too slow to meet the reaction time. Since all cells respond simultaneously, in filamentous blue-green algae, there must be an almost instantaneous information exchange from one end to the other. Therefore, it is more reasonable to assume the involvement of electrical phenomena[43]. In a number of systems, such as those of ciliates[72] and bacteria[21], the utilization of electrical potential changes has been demonstrated.

a) Mitchell's Chemiosmotic Hypothesis

Light-induced electrical potential changes have been demonstrated in both prokaryotic and eukaryotic organisms[32,62,117]. In photosynthetic organisms, thylakoid membranes maintain H^+ gradients equivalent to 200–300 mV[120]. The generation of these potentials can be explained by the chemiosmotic hypothesis[80,81].

The redox components of the photosynthetic electron transport chain are located within the thylakoid membrane (Fig. 10). When the antenna pigments absorb light and funnel the energy to their proper energy trap, an electron transport takes place along the redox chain[59]. Plastoquinone is believed to pick up two protons from the outside each time it is reduced. During the opposite transition plastoquinone releases the protons on the inner side of the membrane[14,116]. The resulting proton gradient ($\Delta\tilde{\mu}H$) causes an electrical potential ($\Delta\psi$) which adds up to the existing resting potential of the cytoplasmic membrane. This explains the observation of light-induced potential changes which have been measured in several photosynthetic organisms[1,2].

b) External Measurement of Light-Induced Potential Changes

A potential gradient between the two ends of a trichome, caused by light-induced potential changes, can be measured by two flat platinum electrodes mounted at a

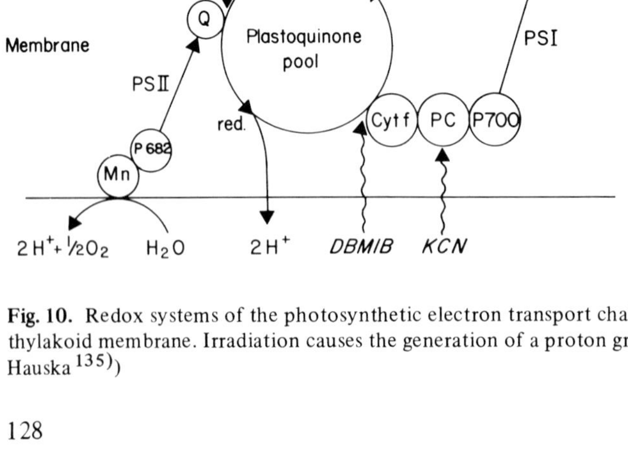

Fig. 10. Redox systems of the photosynthetic electron transport chain incorporated in the thylakoid membrane. Irradiation causes the generation of a proton gradient (after Trebst and Hauska[135])

small distance. When the gap is bridged by perpendicularly oriented trichomes and the system covered by a water agar cube to prevent it from drying (Fig. 11a), a potential can be measured between the two electrodes. A light source irradiates the organisms with a cycle of 45 s dark and 160 s light. This rhythm has been proven microscopically to induce most organisms to undergo repeated photophobic responses. The light-induced electrical potential changes follow this light-dark rhythm with a delay of about 10 s[45]. This delay is most obvious at the end of the light period (Fig. 11b).

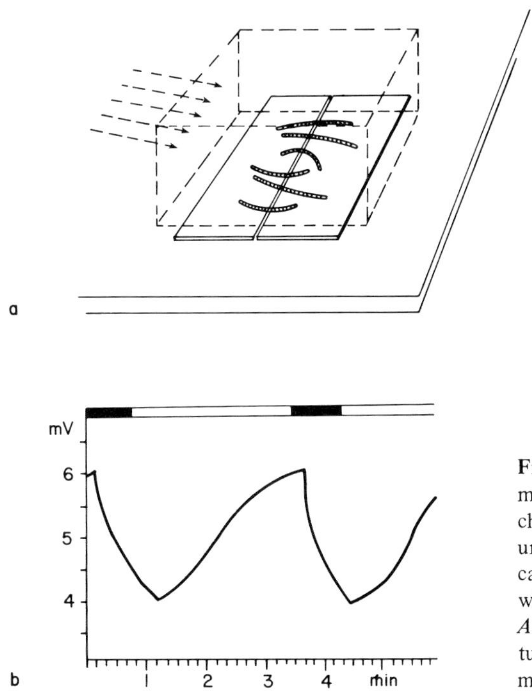

Fig. 11 a. Schematic drawing of instrument to measure light-induced potential changes externally. **b** Externally measured light-induced potential changes caused by a light-dark rhythm of 160 s white light (1000 lx) and 45 s dark. *Abscissa:* time in min; *Ordinate:* amplitude of light-induced potential changes in mV (after Häder[45])

The lag phase coincides well with photophobic reaction time, which has been measured microscopically in single organism experiments. The amplitude of the induced potential changes increases with light intensity to a saturation value of about 10 mV, reached at about 10^4 lx. The action spectrum shows main peaks in the absorption bands of C-phycoerythrin at around 563 nm and chlorophyll a at around 670 nm. Smaller maxima at 620 nm and 710 nm may be due to the activity of C-phycocyanin and P 700 respectively. In the blue region there is a broad band which might represent the combined action of the Soret band of chlorophyll a and the carotenoids. This action spectrum resembles that of both the photophobic response[88] and the photosynthetic CO_2 fixation[105].

The technique described above measures only light-induced potential changes from a whole population of organisms. Therefore, it had to be verified that these changes are based on a light-induced voltage gradient between the two ends of the single trichomes. In order to accomplish this, organisms were inserted from the rear end of a glass micropipette and allowed to move until they had partially emerged from the tip[46]. After microscopic selection for close fitting of the inner tip diameter around the trichome, the front end was dipped into a measurement bath containing the counter electrode, while a Ag/AgCl halfcell was inserted into the rear end of the pipette. The light-induced potential changes followed the light-dark rhythm with a delay about 10 s. The amplitude of up to 10 mV seems to be influenced by the leakage of current between the glass electrode and the trichomes' surface, depending on the electrical resistance between the two ends.

c) Internal Potential Measurements

Potentials generated across the cytoplasmic membrane can be measured by penetrating the cell with microelectrodes attached to a micromanipulator[83,110]. The resting potential of a blue-green alga cell is approximately – 30 to – 40 mV. Measured by a microelectrode with a tip diameter of 0.1 μm, the potential is stable for several hours, indicating that the cells tolerate this treatment and form a good seal around the electrode[46].

A light-dark rhythm with the same duration of light and dark phases as with the external electrodes causes potential changes, which again follow the light-dark oscillations with a lag phase of 10 s. The action spectrum (Fig. 12) shows the involvement of C-phycoerythrin with a peak at 560 nm. The maximum, centered at around 670 nm, is caused by chlorophyll a absorption. Smaller peaks are found at about 620 nm and 710 nm, as is, the case with external measurements, as well as the broad band in the blue region. The broken line represents the action spectrum of photophobic responses measured with a single beam-light trap system[88].

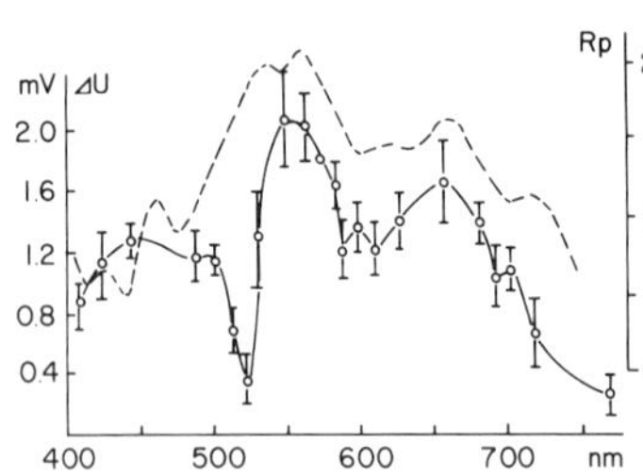

Fig. 12. Action spectrum of light-induced potential changes measured with microelectrodes (after Häder[46]) and action spectrum of photophobic response. *Abscissa:* wavelength in nm; ***Ordinates:*** amplitude of light-induced potential changes in mV and phobic response in relative units respectively (after Nultsch[88])

These findings show that externally measured light-induced potential changes are due to internal potentials generated by the proton pump located in thylakoids. One disadvantage is that it has not yet been possible to measure light-induced potential changes and detect phobic responses simultaneously, because, when penetrated by microelectrodes the organisms cannot move and thus show no photophobic response. Another possible way to detect potential changes is to use fluorescent dyes, which have been used for the same purpose in observing photosynthesis in blue-green algae. Padan and Schuldinger[112] have determined from the distribution of fluorescent amines that illuminating isolated membrane preparations causes a transmembrane pH gradient of about 2 to 3 pH units.

d) Inhibition of Photophobic Responses by External Electric Fields

From the results described above, it may be concluded that a potential is generated along the length of a trichome when it undergoes a decrease in light intensity, thereby causing a photophobic reversal. This internally generated voltage can be canceled by the application of an external dc-field[44]. When a light trap is placed within an electrical field, the accumulation of organisms in the trap is inhibited if the direct current field exceeds a value of 2.5 V. From the field length (30 mm) and the average length of the trichomes (ca. 100 μm), the potential can be calculated which is induced into each organism. A voltage of 10 mV is required to inhibit phobic responses. This is the same potential as that measured on the surface of trichomes with external potentials and indicates that photophobic responses are triggered by light-induced potentials of about 10 mV between the two ends of a filament.

Microscopic analysis of the photobehaviour demonstrated that phobic responses are inhibited only in those organisms which move more or less parallel to the electric field lines. In those leaving the light trap perpendicular to the field lines, the phobic responses were not impaired. But since the light trap was not completely closed, no accumulations were formed.

Photophobic responses can also be inhibited by alternating current fields (Fig. 13). Rectangular waves inhibit phobic responses when the peak-to-peak voltage exceeds a certain threshold U_k, provided that a frequency in the mHz or μHz regions is used. Higher frequencies are not as effective[44]. The same has been found for sine and triangular waves, which are less effective than rectangular ones. From the data it can be concluded that a voltage exceeding the threshold U_k has to be applied for a minimal time t_m in order to cancel the phobic response. This time span happens to be equal to the photophobic reaction time, namely about 10 s.

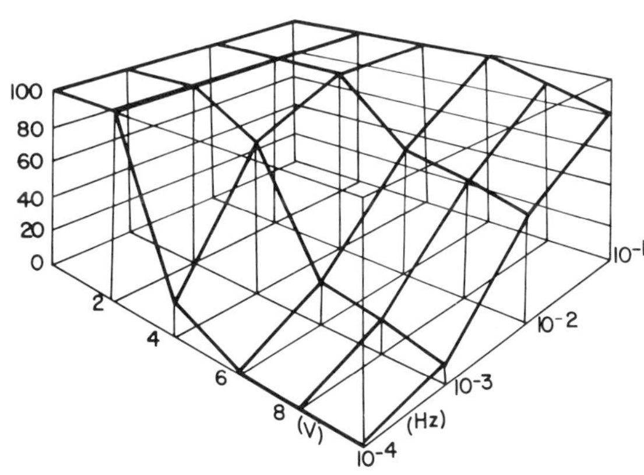

Fig. 13. Three-dimensional representation of the effect of voltage and frequency of a rectangular wave on photophobic accumulations in 1000 lx white light traps. *Left abscissa:* peak-to-peak voltage; *Right abscissa:* frequency in Hz; *Ordinate:* response in % of the uninhibited control (after Häder[44])

Marbach and Mayer[74] found a similar inhibition of phototactic orientation in the flagellate *Chlamydomonas* when they applied an external electric field of 10 V/cm. This voltage introduced a potential of about 10 mV into the cells. Movement was not impaired by the field in either *Chlamydomonas* or *Phormidium*.

e) Effects of Uncouplers and Inhibitors

The inhibitor of the non-cyclic photosynthetic electron transport chain, DCMU, impairs the photophobic response in *Phormidium uncinatum* (see Sect. II.3). According to the Mitchell chemiosmotic hypothesis, an interruption of the electron transport results in a decrease of the proton gradient, thereby causing the light-induced potential to drop. If this assumption is true, then the effect of inhibitors and external fields should be additive, since both impair light-induced potential changes. Figure 14 shows the effect of DCMU on the photophobic response in light traps (dots). Fifty per cent inhibition is obtained in a concentration of about $3 \cdot 10^{-5}$ M, and 10^{-4} M completely inhibits the response[48]. The concentrations needed to block the phobic reaction are higher than those needed to inhibit photosynthetic oxygen evolution[138]. This may be due to the fact that accumulation of white light involves both photosystems, while only photosystem II is inhibited by DCMU[121]. The application of a dc voltage of 2.5 Volt does not impair the phobic response at all, whereas simultaneous application of the external potential enhances the effect of DCMU on the phobic reaction by a factor of $4 \cdot 10^2$ for the 50% inhibition, and by a factor of 10^2 for the complete inhibition respectively. These results indicate that inhibitors of the photosynthetic electron transport chain and the application of an external electrical potential do impair the phobic response by inhibiting the generation of a light-induced potential change.

Another way to reduce membrane potentials is to introduce conducting channels into the membrane. Gramicidin A, for example, acts as an ion channel for monovalent

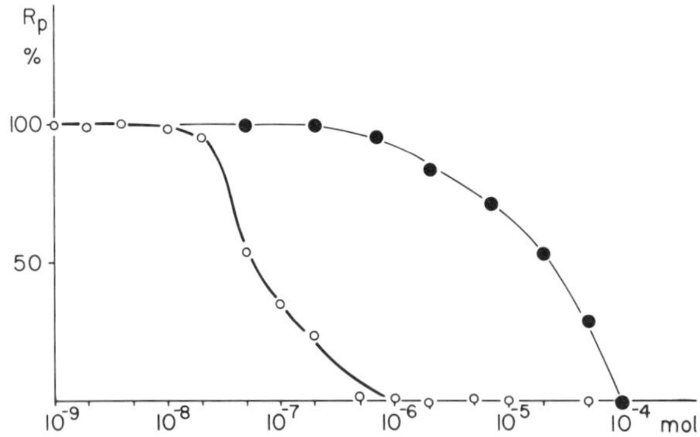

Fig. 14. Inhibition of photophobic responses R_p by DCMU in the absence *(closed circles)* and presence *(open circles)* of an electric dc field of 0.83 V/cm. *Abscissa:* DCMU:concentration in mol; *Ordinate:* response in % of the uninhibited control (after Häder[48])

ions[82,122], while some uncouplers have protonophorous properties[29]. $TPMP^+$ has been demonstrated to penetrate the cytoplasmic membrane of *Halobacterium halobium*, thus canceling the membrane potential[78]. This drug has the same effect on the resting potential in blue-green algae (Fig. 15). The inhibition curve of the light-induced potential change looks almost the same, indicating that $TPMP^+$ penetrates thylakoids as well. The effect of $TPMP^+$ on the photophobic response is similar and can be regarded as evidence that photophobic responses are indeed triggered by light-induced potential changes.

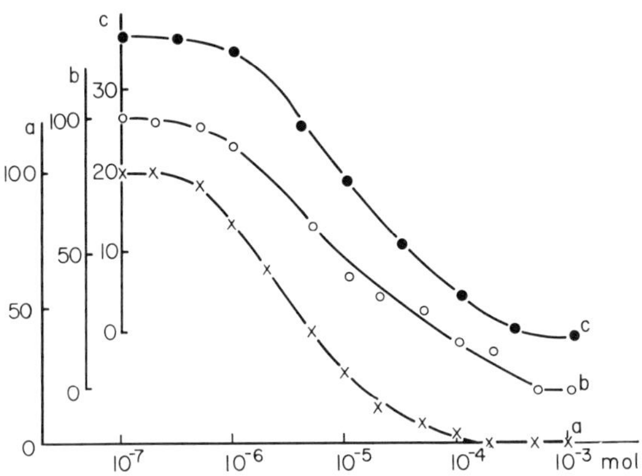

Fig. 15a–c. Effect of $TPMP^+$ on **(a)** light-induced potential changes *(Left ordinate* in % of uninhibited control), **(b)** on photophobic accumulations in light traps (*Center ordinate* in % of uninhibited control) and **(c)** resting potential (*Right ordinate* in mV). *Abscissa:* $TPMP^+$ concentration in mol (after Häder[48])

V. Motor Responses

The mechanism of movement in blue-green algae is still hypothetical (cf. reviews by Walsby[137], Castenholz[12]; Fogg et al.[31]).

The extrusion of slime seems to be unlikely, since it is difficult to imagine the slime stream to swing around each time the organism undergoes a photophobic response. Another hypothesis is based on observing oscillating waves traveling over the surface of the trichome. Halfen and Castenholz[51,52] described microfilaments with a 5 to 8 nm diameter in a position parallel to the outer membrane. Different species of blue-green algae rotate either right or left during their forward movement, and the angle of rotation coincides with that of filament orientation[53,131]. Thus, the mechanism of movement could be explained by the contraction of filaments[9]; but on the other hand, cytochalasin B, which inhibits filament contraction in other systems, does not impair movement in blue-green algae[50].

In Cyanophyceae it is still premature to speculate on the connection between light-induced potential changes and movement on a molecular level. The photosynthetic bacterium *Rhodospirillum* produces membrane potentials upon illumination[22, 70], as well as *Halobacterium*[123], which can be measured by monitoring the fluorescence of cyanine dyes[79] or by observing Rb^{86} or $TPMP^+$ distribution[129]. The high energy intermediate[65], which is used as a chemical transmitter (see above), seems to be converted into a proton gradient and directly utilized for the molecular engine, which drives the bacterial flagellum in *Rhodospirillum rubrum*[35]. A model for the flagellar movement is based on observing a proton flux from outside between the cell wall and the flagellar rod driving the M-ring into a rotary movement[64]. Thus, it is easy to imagine how a variation in the proton gradient causes a change in movement. Modulating the potential by changes in light intensity directly influences the motor response. Whether or not a similar mechanism applies to the sensory transduction in photophobic responses of blue-green algae remains to be investigated.

VI. Development

Vernotte[136] observed absorption spectra for two types of chromophores (f- and s-chromophores) in C-phycocyanin of *Phormidium*. The difference spectrum between the two coincides with the absorbance change of a phycochrome. According to the low ΔOD, this change takes place in only about 5% of the total amount of C-phycocyanin. Björn and Björn[5] developed a model to explain these experimental results (Fig. 16). The protein component of phycocyanin is supposed to possess two binding sites, to one of which a chromophore (phycocyanobilin) is attached. This single chromophore can be shifted between the two binding sites by irradiation with green or red light and results in the observed absorption change. The final step is the additional binding of a second chromophore. Glazer and Fang[36] even suggest that three chromophores are bound per monomer. The absorption spectra of the s- and f-chromophore phycobilins coincide with the two absorption forms of phycochrome a[5], suggesting the identity of phycochrome and a precursor of C-phycocyanin.

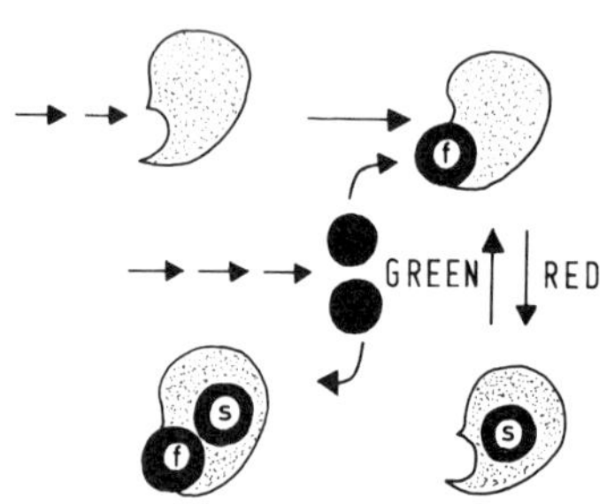

Fig. 16. Model for the synthesis of phycocyanin. In the first step only one chromophore is attached to the protein and can be shifted between the two binding sites by green or red irradiation. This is terminated by the attachment of a second chromophore (after Björn and Björn[5])

Although the action spectra for the induction of the filamentous growth in *Nostoc* are shown to coincide with the difference spectrum between the two absorption forms of phycochrome c (see above), no molecular mechanism for the transduction has yet been suggested.

DBMIB
2,5-dibromo-3-methyl-6-isopropyl-p-benzoquinone

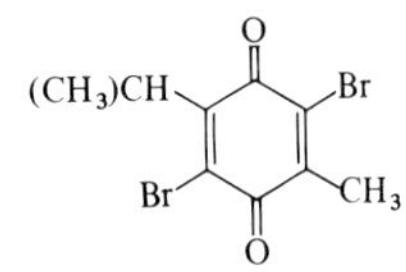

$TPMP^+$
triphenyl-methyl-phosphonium as bromide

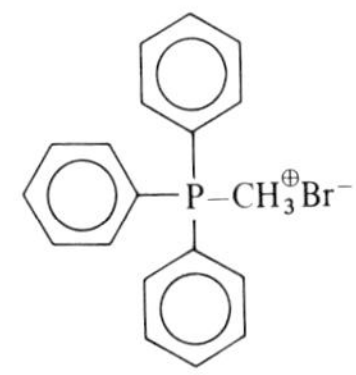

DCMU
3-(3′,4′-dichlorophenyl)-1,1-dimethylurea

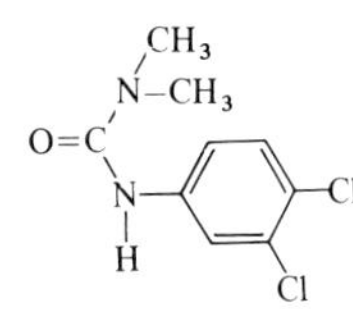

DNP
2,4-dinitrophenol

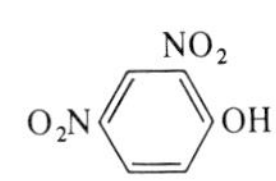

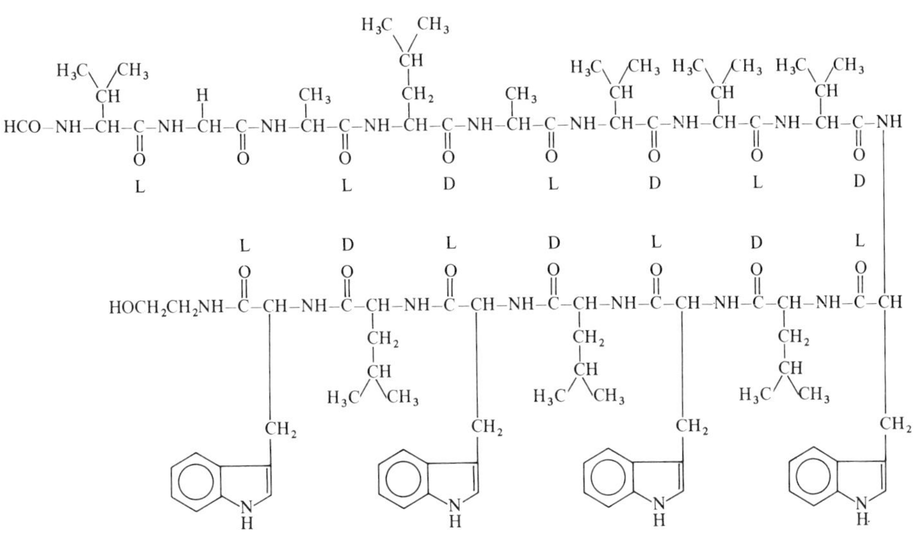

valine-gramicidin A

VII. References

1. Ausländer, W., Junge, W.: FEBS Lett. *59*, 310 (1975)
2. Bentrup, F.-W.: Ber. dtsch. Bot. Ges. *87*, 215 (1974)
3. Björn, G.S.: Physiol. Plant. *42*, 321 (1978)
4. Björn, G.S.: Physiol. Plant. *46*, 281 (1979)
5. Björn, G.S., Björn, L.O.: Physiol. Plant. *36*, 297 (1976)
6. Björn, G.S., Björn, L.O.: ibid. *43*, 195 (1978)
7. Bogorad, L.: Ann. Rev. Plant. Physiol. *26*, 369 (1975)
8. Bovee, E.C., Jahn, Th.L.: J. Theor. Biol. *35*, 259 (1972)
9. Bray, D.: Nature *273*, 265 (1978)
10. Butler, W.L.: Chlorophyll fluorescence: A probe for electron transfer and energy transfer. In: Encyclopedia of Plant Physiology, New Series, Vol. 5 Photosynthesis I. (Trebst, A., Avron, M., eds.). Berlin, Heidelberg, New York: Springer 1977, pp. 149–167
11. Caserta, G., Cervigni, T.: J. Theor. Biol. *41*, 127 (1973)
12. Castenholz, R.W.: Movements. In: Botanical Monographs, Vol. 9 (Carr, N.G., Whitton, B.A., eds.). Oxford: Blackwell 1973, pp. 320–339
13. Chapmann, D.J.: Biliproteins and bile pigments. In: The Biology of Blue-Green Algae (Carr, N.G., Whitton, B.A., eds.). Oxford: Blackwell 1973, pp. 162–185
14. Chow, W.S., Thorne, S.W., Boardman, N.K.: The movement of protons during energy transduction in the chloroplast thylakoid membrane. In: Light Transducing Membranes (Deamer, D.W., ed.). New York, San Francisco, London: Academic Press 1978, pp. 253–268
15. Clayton, R.K.: Arch. Mikrobiol. *19*, 107 (1953)
16. Clayton, R.K.: ibid. *19*, 125 (1953)
17. Clayton, R.K.: Phototaxis of purple bacteria. In: Encyclopedia of Plant Physiology (Ruhland, W., ed.). Vol. 17/1. Berlin, Göttingen, Heidelberg: Springer 1959, pp. 371–387
18. Codd, G.A., Cossar, J.D.: Biochem. Biophys. Res. Comm. *83*, 342 (1978)
19. Diakoff, St., Scheibe, J.: Plant Physiol. *51*, 382 (1973)
20. Dietrich, W.E., Thornber, J.P.: Biochim. Biophys. Acta *245*, 482 (1971)
21. Doetsch, R.N.: J. Theor. Biol. *35*, 55 (1972)
22. Drachev, L.A. et al.: Biochim. Biophys. Acta *440*, 637 (1976)
23. Drews, G.: Structure and development of the membrane system of photosynthetic bacteria. In: Current Topics in Bioenergetics. Vol. 8. New York, San Francisco, London: Academic Press 1978, pp. 161–207
24. Drews, G., Jaeger, K.: Nature *199*, 1112 (1963)
25. Duysens, L.N.M.: Transfer of excitation energy in photosynthesis. Thesis, Utrecht 1952
26. Engelmann, Th.W.: Pflügers Arch. ges. Physiol. *29*, 387 (1882)
27. Engelmann, Th.W.: ibid. *30*, 95 (1883)
28. Engelmann, Th.W.: Bot. Ztg. *46*, 661 (1888)
29. Felle, H., Bentrup, F.W.: Biochim. Biophys. Acta *464*, 179 (1977)
30. Firsow, N.N., Drews, G.: Arch. Microbiol. *115*, 299 (1977)
31. Fogg, G.E. et al.: The blue-green algae. London, New York: Academic Press 1973
32. Fujii, S., Shimmen, T., Tazawa, M.: Plant Cell Physiol. *19*, 573 (1978)
33. Fujita, Y., Hattori, A.: Plant and Cell Physiol. *3*, 209 (1962)
34. Gingras, G.: A comparative review of photochemical reaction center preparations from photosynthetic bacteria. In: The Photosynthetic Bacteria (Clayton, R.K., Sistrom, W.R., eds.). New York: Plenum Press 1978, pp. 119–131
35. Glagolev, A.N., Skulachev, V.P.: Nature *272*, 280 (1978)
36. Glazer, A.N., Fang, S.: J. Biol. Chem. *248*, 659 (1973)
37. Göbel, F.: Quantum efficiencies of growth. In: The Photosynthetic Bacteria (Clayton, R.K., Sistrom, W.R., eds.). New York: Plenum Press 1978, pp. 907–925
38. Goodwin, T.W.: Carotenoids and biliproteins. In: Algal Physiology and Biochemistry. Vol. 10 (Stewart, W.D.P., ed.). Oxford, London, Edinburgh, Melbourne: Blackwell Scientific 1974, pp. 176–205

39. Häder, D.-P.: Untersuchungen zur Photo-phobotaxis von Phormidium uncinatum. Thesis, Marburg 1973
40. Häder, D.-P.: Arch. Microbiol. *96*, 255 (1974)
41. Häder, D.-P.: ibid. *103*, 169 (1975)
42. Häder, D.-P.: Z. Pflanzenphysiol. *78*, 173 (1976)
43. Häder, D.-P.: Speculations about sensory transduction. In: Research in Photobiology (Castellani, A., ed.). New York, London: Plenum Press 1977, pp. 95–102
44. Häder, D.-P.: Arch. Microbiol. *114*, 83 (1977)
45. Häder, D.-P.: ibid. *118*, 115 (1978)
46. Häder, D.-P.: ibid. *119*, 75 (1978)
47. Häder, D.-P.: Photomovement. In: Encyclopedia of Plant Physiology. New Series, Vol. 7 (Haupt, W., Feinleib, M.E., eds.). 1979, pp. 268–309
48. Häder, D.-P.: Arch. Microbiol. *120*, 57 (1979)
49. Häder, D.-P., Nultsch, W.: Photochem. Photobiol. *18*, 311 (1973)
50. Halfen, L.N.: J. Phycol. *9*, 248 (1973)
51. Halfen, L.N., Castenholz, R.W.: Nature *225*, 1163 (1970)
52. Halfen, L.N., Castenholz, R.W.: J. Phycol. *7*, 133 (1971)
53. Halfen, L.N., Castenholz, R.W.: J. Phycol. *7*, 258 (1971)
54. Harayama, S., Iino, T.: J. Bacteriol. *131*, 34 (1977)
55. Harayama, S., Iino, T.: Photochem. Photobiol. *25*, 571 (1977)
56. Harold, F.M.: Ann. Rev. Microbiol. *31*, 181 (1977)
57. Haupt, W.: Int. Rev. Cytol. *19*, 267 (1966)
58. Haxo, F.T., Norris, P.S.: Biol. Bull. *105*, 374 (1953)
59. Hinkle, P.C., McCarty, R.E.: Sci. Am. March *238,* 104 (1978)
60. Hoch, G.E.: P-700. In: Encyclopedia of Plant Physiology, New Series, Vol. 5, Photosynthesis I (Trebst, A., Avron, M., eds.). Berlin, Heidelberg, New York: Springer 1977, pp. 136–148
61. Koka, P., Song, P.-S.: Biochim. Biophys. Acta *495*, 220 (1977)
62. Konishi, T., Packer, L.: FEBS Lett. *89*, 333 (1978)
63. Krogmann, D.W.: Blue-green algae. In: Encyclopedia of Plant Physiology, New Series, Vol. 5, Photosynthesis I (Trebst, A., Avron, M., eds.). Berlin, Heidelberg, New York: Springer 1977, pp. 625–636
64. Läuger, P.: Nature *268*, 360 (1977)
65. Larsen, S.H. et al.: Proc. Natl. Acad. Sci. U.S.A. *71*, 1239 (1974)
66. Lazaroff, N.: J. Phycol. *2*, 7 (1966)
67. Lazaroff, N.: Photomorphogenesis and nostocacean development. In: Botanical Monographs, Vol. 9. The Biology of Blue-green Algae (Carr, N.G., Whitton, B.A., eds.). Oxford, London, Edinburgh, Melbourne: Blackwell 1973, pp. 279–319
68. Lazaroff, N., Schiff, J.: Science *137*, 603 (1962)
69. Lee, A.G., Fitzsimons, J.T.R.: J. Gen. Microbiol. *93*, 346 (1976)
70. Leiser, M., Gromet-Elhanan, Z.: Arch. Biochem. Biophys. *178*, 79 (1977)
71. Links, J.: An hypothesis for the mechanism of (phobo-) chemotaxis. II. The carotenoids, steroids and fatty acids of Polytoma uvella. Thesis, Leiden, 1955
72. Machemer, H.: J. Exp. Biol. *75*, 427 (1976)
73. Manten, A.: Phototaxis, phototropism and photosynthesis in purple bacteria and blue-green algae. Thesis, Utrecht, 1948
74. Marbach, I., Mayer, A.M.: Israel J. Bot. *20*, 96 (1971)
75. Mauzerall, D.: Porphyrins, chlorophyll and photosynthesis. In: Encyclopedia of Plant Physiology, New Series, Vol. 5, Photosynthesis I (Trebst, A., Avron, M., eds.). Berlin, Heidelberg, New York: Springer 1977, pp. 117–124
76. Mauzerall, D.: Bacteriochlorophyll and photosynthetic evolution. In: The Photosynthetic Bacteria (Clayton, R.K., Sistrom, W.R., eds.). New York: Plenum Press 1978, pp. 223–231
77. Meeks, J.K.: Chlorophylls. In: Algal Physiology and Biochemistry. Botanical Monographs, Vol. 10 Stewart, W.D.P., ed.). Oxford: Blackwell 1974, pp. 161–175
78. Michel, H., Oesterhelt, D.: FEBS Lett. *65*, 175 (1976)
79. Miller, J.B., Koshland, D.E. Jr.: Nature *272*, 83 (1978)

80. Mitchell, P.: Biol. Rev. *41*, 445 (1966)
81. Mitchell, P.: TIBS *3*, N58 (1978)
82. Neher, E., Sandblom, J., Eisenman, G.: J. Membr. Biol. *40*, 97 (1978)
83. Nelles, A.: Biochem. Physiol. Pflanzen *171*, 147 (1977)
84. Nichols, B.W.: Lipid composition and metabolism. In: The Biology of Blue-green Algae (Carr, N.G., Whitton, B.A., eds.), Botanical Monographs, Vol. 9. Oxford, London, Edinburgh, Melbourne: Blackwell 1973, pp. 144–161
85. Nielsen, N.C., Smillie, R.M.: Arch. Biochem. Biophys. *186*, 52 (1978)
86. Nultsch, W.: Planta *56*, 632 (1961)
87. Nultsch, W.: ibid. *57*, 613 (1962)
88. Nultsch, W.: ibid. *58*, 647 (1962)
89. Nultsch, W.: Ber. dtsch. Bot. Ges. *75*, 443 (1962)
90. Nultsch, W.: Photochem. Photobiol. *4*, 613 (1965)
91. Nultsch, W.: Arch. Microbiol. *55*, 187 (1966)
92. Nultsch, W.: Z. Pflanzenphysiol. *56*, 1 (1967)
93. Nultsch, W.: Arch. Microbiol. *63*, 292 (1968)
94. Nultsch, W.: Photochem. Photobiol. *10*, 119 (1969)
95. Nultsch, W.: Photomotion of microorganisms and its interaction with photosynthesis. In: Photobiology of Microorganisms (Halldal. P., ed.). London, New York, Sidney, Toronto: Wiley 1970, pp. 213–251
96. Nultsch, W.: Relation between photomotion and photosynthesis. In: Primary Molecular Events in Photobiology (Checcucci, A., Weale, R.A., eds.). Amsterdam, London, New York: Elsevier 1973, pp. 245–273
97. Nultsch, W.: Phototaxis and photokinesis in bacteria and blue-green algae. In: Behaviour of Microorganisms (Pérez-Miravete, A., eds.). London, New York: Plenum Press 1973, pp.70–81
98. Nultsch, W.: Movements. In: Algal Physiology and Biochemistry. Botanical Monographs Vol. 10 (Stewart, W.D.P., ed.). Oxford, London, Edinburgh, Melbourne: Blackwell 1974, pp. 864–893
99. Nultsch, W.: Der Einfluß des Lichtes auf die Bewegung phototropher Mikroorganismen. I. Photokinesis. In: Abhandlungen Marb. Gel. Ges. Jg. 1972 Nr. 2. München: Fink 1974, pp. 141–213
100. Nultsch, W.: Phototaxis and photokinesis. In: Primitive Sensory and Communication Systems: The Taxes and Tropisms of Micro-organisms and Cells (Carlile, M.J., ed.). London, New York, San Francisco: Academic Press 1975, pp. 29–90
101. Nultsch, W., Häder, D.-P.: Ber. dtsch. Bot. Ges. *87*, 83 (1974)
102. Nultsch, W., Häder, D.-P.: Photochem. Photobiol. *29*, 423 (1979)
103. Nultsch, W., Hellmann, W.: Arch. Mikrobiol. *82*, 76 (1972)
104. Nultsch, W., Jeeji-Bai, N.: Z. Pflanzenphysiol. *54*, 84 (1966)
105. Nultsch, W., Richter, G.: Arch. Mikrobiol. *47*, 207 (1963)
106. Nultsch, W., Schuchart, H., Dillenburger, M.: Arch. Microbiol. *122*, 207 (1979)
107. O'Carra, P., O'hEocha, C.: Algal biliproteins and phycobilins. In: Chemistry and Biochemistry of Plant Pigments. Vol. 1 (Goodwin, T.W., ed.). London: Academic Press 1976, pp. 328–376
108. Ohki, K., Fujita, Y.: Plant & Cell Physiol. *19*, 7 (1978)
109. Ohki, K., Fujita, Y.: ibid. *20*, 483 (1979)
110. Okada, Y., Inouye, A.: Biophys. Struct. Mechan. *2*, 31 (1976)
111. Olson, J.M.: Bacteriochlorophyll a-proteins from green bacteria. In: The Photosynthetic Bacteria (Clayton, R.K., Sistrom, W.R., eds.). New York: Plenum Press 1978, pp. 161–178
112. Padan, E., Schuldinger, S.: J. Biol. Chem. *253*, 3281 (1978)
113. Pfennig, N.: General physiology and ecology of photosynthetic bacteria. In: The Photosynthetic Bacteria (Clayton, R.K., Sistrom, W.R., eds.). New York: Plenum Press 1978, pp. 3–18
114. Pfennig, N., Trüper, H.G.: The phototrophic bacteria. In: Bergey's Manual of Determinative Bacteriology (Buchanan, R.E., Gibbons, N.E., eds.). 8th Ed. Baltimore, MD: Williams & Wilkins 1974, pp. 24–64

115. Piccinni, E., Omodeo, P.: Boll. Zool. *42*, 57 (1975)
116. Poole, R.J.: Ann. Rev. Plant Physiol. *29*, 437 (1978)
117. Racusen, R.H.: Planta *132*, 25 (1976)
118. Rüdiger, W.: Photochemistry of phycobilins. In: Biophysics of Photoreceptors and Photobehaviour of Microorganisms (Colombetti, G., ed.). Pisa: Litofelici 1975, pp. 184–197
119. Rüdiger, W.: Ber. dtsch. Bot. Ges. (1980) in press
120. Ryrie, I.J.: The contribution of reconstituted proteinlipid vesicles to the elucidation of energy transduction mechanisms in membranes. In: Light Transducing Membranes (Deamer, D.W., ed.). New York, San Francisco, London: Academic Press 1978, pp. 201–213
121. Satoh, K., Strasser, R., Butler, W.L.: Biochim. Biophys. Acta *440*, 337 (1976)
122. Schagina, L.V., Grinfeldt, A.E., Lev, A.A.: Nature *273*, 243 (1978)
123. Schulten, K., Tavan, P.: Nature *272*, 85 (1978)
124. Schumacher, A., Drews, G.: Biochim. Biophys. Acta *50*, 183 (1978)
125. Siggel, U.: Bioelectrochem. Bioenergetics *3*, 302 (1976)
126. Sistrom, W.R.: Phototaxis and Chemotaxis. In: The Photosynthetic Bacteria. (Clayton, R.K., Sistrom, W.R., eds.). New York, London: Plenum Press 1978, pp. 899–905
127. Song, P.-S.: J. Korean Agric. Chem. Soc. *20*, 10 (1977)
128. Song, P.-S. et al.: Biochem. *15*, 4422 (1976)
129. Sprott, G.D., Usher, J.R.: Can. J. Microbiol. *23*, 1683 (1977)
130. Stanier, R.Y.: Symp. Soc. Gen. Microbiol. *24*, 219 (1974)
131. Thomas, E.A.: Schweiz. Z. Hydrologie *32*, 523 (1970)
132. Thornber, J.P., Trosper, T.L., Strouse, C.E.: Bacteriochlorophyll in vivo: relationship of spectral forms to specific membrane components. In: The Photosynthetic Bacteria (Clayton, R.K., Sistrom, W.R., eds.). New York: Plenum Press 1978, pp. 133–160
133. Throm, G.: Arch. Protistenk. *110*, 313 (1968)
134. Trebst, A., Harth, E., Draber, W.: Z. Naturforsch. *25b*, 1157 (1970)
135. Trebst, A., Hauska, G.: Naturwissenschaften *61*, 308 (1974)
136. Vernotte, C.: Photochem. Photobiol. *14*, 163 (1971)
137. Walsby, A.E.: Protoplasma *65*, 223 (1968)
138. Yamashita, T., Tomita, G.: Plant Cell Physiol. *17*, 571 (1976)
139. Zuber, H.: Ber. dtsch. Bot. Ges. (1980) in press

Author-Index Volumes 1—41